MATHEMATICAL GEMS

FROM ELEMENTARY COMBINATORICS, NUMBER THEORY, AND GEOMETRY

By
ROSS HONSBERGER

THE
DOLCIANI MATHEMATICAL EXPOSITIONS

Published by
THE MATHEMATICAL ASSOCIATION OF AMERICA

The Dolciani Mathematical Expositions

NUMBER ONE

MATHEMATICAL GEMS

FROM ELEMENTARY COMBINATORICS, NUMBER THEORY, AND GEOMETRY

By
ROSS HONSBERGER
University of Waterloo

Published and Distributed by
THE MATHEMATICAL ASSOCIATION OF AMERICA

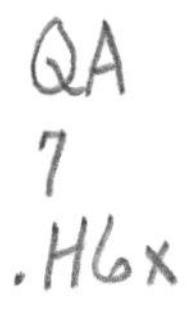

Library of Congress Catalog Card Number 73-89661

Complete Set ISBN 0-88385-300-0
Vol. 1 ISBN 0-88385-301-9

Printed in the United States of America

Current printing (last digit):

10 9 8 7 6 5 4 3 2 1

FOREWORD

The DOLCIANI MATHEMATICAL EXPOSITIONS series of the Mathematical Association of America came into being through a fortuitous conjunction of circumstances.

Professor Mary P. Dolciani, of Hunter College of the City University of New York, herself an exceptionally talented and enthusiastic teacher and writer, had been contemplating ways of furthering the ideal of excellence in mathematical exposition.

At the same time, the Association had come into possession of the manuscript for the present volume, a collection of essays which seemed not to fit precisely into any of the existing Association series, and yet which obviously merited publication because of its interesting content and lucid expository style.

It was only natural, then, that Professor Dolciani should elect to implement her objective by establishing a revolving fund to initiate this series of MATHEMATICAL EXPOSITIONS.

The books in the series will be selected both for clear, informal style and for stimulating mathematical content. It is anticipated that each will have an ample supply of exercises, many with accompanying solutions. Hence they should provide valuable enrichment material, especially for survey courses. It is contemplated that the series will be reasonably comprehensible to the mathematically talented high school student, and yet also challenging to the more advanced mathematician.

The Mathematical Association of America, for its part, is delighted to accept the generous gift establishing this series from one who has served the Association with distinction, both as a member of the Committee on Publications and as a member of the Board of Governors. It is with genuine pleasure that the Board has chosen to name the series in her honor.

Edwin F. Beckenbach
Chairman, Committee on Publications
Mathematical Association of America

PREFACE

This book presents thirteen topics from elementary mathematics. Each chapter has a simple theme and contains at least one mathematical gem. They are presented in the hope that you will come to know the thrill of some of the best discoveries at the elementary level.

The comparison of mathematics and music is often particularly apt. The most attractive music is spoiled by a bad performance. So it is that many an admirable mathematical thought languishes amid the colorless rigor of a formal exposition. The present work seeks to charm you with some wonderful pieces of mathematics. I have endeavored to relate them so that they can be followed and enjoyed without frustration.

No matter what the level, the reader must possess an appropriate background in order to proceed comfortably. To say that a topic is elementary is not necessarily to imply that it is easy or simple. For much of this book, the reader need have little technical knowledge beyond high school mathematics. It is assumed that he is familiar also with the binomial theorem, mathematical induction, and arithmetic congruences. However it is not expected that the book will be easy reading for many graduates straight out of high school. Some degree of mathematical maturity is presumed, and upon occasion one is required to do some rather careful thinking. It is hoped that the book will be of special interest to high school mathematics teachers and to prospective teachers.

The essays are virtually independent, and they may be read in any order. Exercises, references, and suggestions for further reading conclude each topic. The reader is encouraged to consider the exercises carefully, for among them are to be found some splendid problems.

I should like to take this opportunity to thank Dr. Ralph Boas for a review of the work which resulted in many major changes. I should also like to thank four colleagues: Dr. U. S. R. Murty for assistance with the chapter on the Kozyrev-Grinberg Theory of Hamiltonian Circuits, and Fred Miller, Dr. Paul Schellenberg, and Ed Anderson for critically reading parts of the manuscript. Finally, I thank Professor Paul Erdös for pointing out several inaccuracies and for making various helpful comments and suggestions.

ROSS HONSBERGER

University of Waterloo

CONTENTS

CHAPTER 1

AN OLD CHINESE THEOREM AND PIERRE DE FERMAT

1. Fermat's Simple Theorem. In a letter to Bernard Frenicle de Bessy in 1640, the great French number theorist Pierre de Fermat stated, without proof, the following theorem:

If p is a prime number, then for every integer a, the number $a^p - a$ is divisible by p.[1]

For example, because 23 is a prime number, $a^{23} - a$ is divisible by 23, no matter what integer (positive, negative, or zero) is substituted for a. Fermat claimed he had a proof, but he did not include it in the letter. The first published proof is due to Euler, almost a hundred years later, although there is evidence that Leibniz proved it around 1683. In view of this, it is almost too much to hope for an easy proof. Therefore it is surprising to learn that it can be proved in a straightforward manner using nothing more than the binomial theorem and mathematical induction. Since the proof is quite attractive and the theorem so fundamental, let us pause briefly in order to work through it.

Let p denote a prime number. For $a = 0$ and for $a = 1$, the value of $a^p - a$ is zero, which is divisible by p. Let us consider positive values of a. We proceed by induction.

Suppose the hypothesis holds for some positive integer a, that is,

[1] Actually he stated the equivalent theorem: *If p is prime, then p divides $a^{p-1} - 1$ for every integer a that is relatively prime to p.*

that p does divide $a^p - a$. We show that the hypothesis holds also for the value $a + 1$, that is, that

$$(a + 1)^p - (a + 1) \qquad \text{is divisible by } p.$$

The binomial theorem gives

$$(a + 1)^p = a^p + \binom{p}{1}a^{p-1} + \binom{p}{2}a^{p-2} + \cdots + \binom{p}{p-1}a + 1.$$

Transposing yields

$$(1) \quad (a + 1)^p - a^p - 1 = \binom{p}{1}a^{p-1} + \binom{p}{2}a^{p-2} + \cdots + \binom{p}{p-1}a.$$

On the right-hand side of this equation, we find that each coefficient $\binom{p}{k}$, $k = 1, 2, \ldots, p - 1$, is divisible by the prime p. We obtain this from the very definition of $\binom{p}{k}$, namely,

$$\binom{p}{k}\cdot k! = p(p - 1)(p - 2)\cdots(p - k + 1).$$

Here p divides the right-hand side, so it must also divide the left-hand side. But for all the coefficients in question, the value of k is less than p. Thus the prime factor p does not occur in the product $k!$ Accordingly, p must divide $\binom{p}{k}$.

Because p divides every coefficient on the right-hand side of equation (1) above, it must divide the entire right-hand side, and consequently it divides the left-hand side, namely $(a + 1)^p - a^p - 1$. Combining this with the induction hypothesis, that p divides $a^p - a$, we see that p divides the sum

$$[(a + 1)^p - a^p - 1] + [a^p - a] = (a + 1)^p - (a + 1),$$

as claimed. Thus by induction the hypothesis is valid for all positive integers a.

We complete the proof by considering negative values of a. Let us denote the negative value in question by $-a$; then a, itself, will denote a positive integer. If the prime p is 2, then we have

$$(-a)^p - (-a) = (-a)^2 - (-a) = a^2 + a = a(a+1).$$

Here the factors a and $a+1$ are consecutive integers. Thus one of them is even and their product is divisible by 2. If p is an odd prime, we have

$$(-a)^p - (-a) = -a^p + a = -(a^p - a).$$

Since a is positive, we know already that p divides $a^p - a$. Thus it also divides $-(a^p - a)$. (QED)

2. An Old Chinese Theorem. For $a = 2$, Fermat's theorem tells us that $2^p - 2$ is divisible by p if p is a prime number. It is natural to ask about the converse of this:

If a positive integer $n > 1$ divides $2^n - 2$, is n a prime number?

The Chinese checked this out for many values of n and came to the conclusion 2500 years ago that the answer is "yes." And modern calculations have confirmed that for $1 < n < 300$ the only values of n which divide $2^n - 2$ are prime numbers.

It is not difficult to show, however, that the old Chinese theorem is wrong. The number $n = 341$ provides a counterexample. Being 11 times 31, 341 is not prime, but it does divide $2^{341} - 2$. The divisibility follows easily from the well-known result that $x^m - y^m$ has the factor $x - y$ for any positive integer m.

We have

$$2^{341} - 2 = 2(2^{340} - 1) = 2[(2^{10})^{34} - 1^{34}] = 2[(2^{10} - 1)(\cdots)]$$

$$= 2[(1023)(\cdots)] = 2[3(341)(\cdots)].$$

Integers n like 341, which are not prime numbers, but which do divide $2^n - 2$, are called **pseudoprimes**.

Now several questions arise in connection with pseudoprimes. Is 341 the only pseudoprime? Is there an infinity of them? It turns out that for every odd pseudoprime there is a greater odd pseudoprime (see exercise 6). Thus the single odd pseudoprime 341 gives rise to an infinity of odd pseudoprimes.

Next it was wondered whether or not there exists an even pseudoprime. It was only in 1950 that the American D. H. Lehmer discovered the pseudoprime 161038. While its discovery was difficult, it is easy to see that 161038 is a pseudoprime. The prime decomposition of 161038 is $2 \cdot 73 \cdot 1103$. Since

$$2^{161038} - 2 = 2(2^{161037} - 1),$$

we need only show that 73 and 1103 divide the number $2^{161037} - 1$.

Now 161037 has the prime decomposition $3^2 \cdot 29 \cdot 617$. Thus

$$2^{161037} - 1 = (2^9)^{29 \cdot 617} - 1^{29 \cdot 617} = (2^9 - 1)(\cdots) = (511)(\cdots)$$
$$= 7(73)(\cdots),$$

showing that it is divisible by 73. Similarly we have

$$2^{161037} - 1 = (2^{29})^{9 \cdot 617} - 1^{9 \cdot 617} = (2^{29} - 1)(\cdots)$$
$$= 1103(486737)(\cdots),$$

showing that 1103 is also a divisor. Since 73 and 1103 are both primes, the divisibility of the one does not affect the divisibility of the other; they both divide $2^{161037} - 1$. Thus 161038 is an even pseudoprime.

In 1951, N. G. W. H. Beeger of Amsterdam proved that there is also an infinity of even pseudoprimes.

3. The Ultimate In Pseudoprimes. A composite number n which divides $2^n - 2$ is a pseudoprime. A composite number n which divides $3^n - 3$, or $4^n - 4$, or etc. . . ., strikes us as sharing in the property of pseudoprimality. A composite number n which divides $2^n - 2$, *and* $3^n - 3$, *and* $4^n - 4$, *and* . . ., *and* $a^n - a$, *and* . . ., for *every* integer a, even the negative integers, is certainly the ultimate in this regard, and is called an **absolute pseudoprime.**

Now you would naturally wonder whether or not any absolute pseudoprimes exist. The answer is "yes." The smallest one is 561. That is to say, 561 is a composite number and $a^{561} - a$ is divisible by 561 no matter what integer is put for a. This is not difficult to show. It follows directly from Fermat's Simple Theorem.

The prime decomposition of 561 is $3 \cdot 11 \cdot 17$. We need to show that $a^{561} - a$ is divisible by each of these primes. We have

$$a^{561} - a = a(a^{560} - 1) = a[(a^{10})^{56} - 1^{56}] = a[(a^{10} - 1)(\cdots)$$
$$= (a^{11} - a)(\cdots).$$

But $a^{11} - a$ is divisible by 11, by Fermat's theorem, because 11 is a prime number. Thus 11 divides $a^{561} - a$. Similarly 3 and 17 are also shown to be divisors.

A few other absolute pseudoprimes are

(i) $2821 = 7 \cdot 13 \cdot 31$ (ii) $10585 = 5 \cdot 29 \cdot 73$ (iii) $15841 = 7 \cdot 31 \cdot 73$.

It is unknown whether or not there exists an infinity of absolute pseudoprimes.

4. The Fermat Numbers. The numbers $F_n = 2^{(2^n)} + 1$, $n = 0, 1, 2, \ldots$, are called **Fermat numbers**. The first few are

$$F_0 = 3, \quad F_1 = 5, \quad F_2 = 17, \quad F_3 = 257, \quad F_4 = 65537,$$
$$F_5 = 2^{(2^5)} + 1 = 2^{32} + 1 = 4294967297.$$

From the time of Fermat it has been known that F_0, F_1, F_2, F_3, and F_4 are prime numbers. It is not easy to classify as prime or composite any of the greater Fermat numbers. Fermat claimed, however, that all the F_n are prime numbers. He admitted that he had no actual proof of this, but he stated flatly that he was certain of it. It may well be that Fermat knew about and believed the Old Chinese Theorem, for it is not difficult to show that every F_n does divide $2^{F_n} - 2$. We may proceed to show this as follows.

Since the first few Fermat numbers are known to be primes, in these cases the conclusion follows immediately from Fermat's Simple Theorem. We need only consider $n = 5, 6, 7, \ldots$. For every n in this range, it follows easily by induction that $n + 1 < 2^n$. This implies that 2^{n+1} divides $2^{(2^n)}$. Thus, for some integer k, we have $2^{(2^n)} = 2^{n+1} \cdot k$. Consequently,

$$\begin{aligned} 2^{F_n} - 2 &= 2^{(2(2^n)+1)} - 2 = 2[2^{(2(2^n))} - 1] = 2[2^{(2^{n+1} \cdot k)} - 1] \\ &= 2[(2^{(2^{n+1})})^k - 1^k] = 2[(2^{(2^{n+1})} - 1)(\cdots)] \\ &= 2[((2^{(2^n)})^2 - 1^2)(\cdots)] = 2[(2^{(2^n)} + 1)(2^{(2^n)} - 1)(\cdots)] \\ &= 2[(F_n)(2^{(2^n)} - 1)(\cdots)]. \end{aligned}$$

Thus it is understandable that Fermat claimed all F_n to be primes.

At present, the nature of 49 Fermat numbers have been determined, and except for those up to F_4 every one of them was found to be composite. As early as 1732 it was known that Fermat was wrong. In that year Euler showed that F_5 is divisible by 641. Although we shall not prove it here, the following theorem of Lucas belongs to elementary number theory:

Every divisor of F_n is of the form $2^{n+2} \cdot k + 1$.

Thus every divisor of F_n occurs in the arithmetic progression

$$1,\ 2^{n+2} + 1,\ 2 \cdot 2^{n+2} + 1,\ 3 \cdot 2^{n+2} + 1, \ldots.$$

For given n, then, we can work out terms of this progression and check to see if any is a divisor of F_n. For $n = 5$, we obtain the sequence 1, 129, 257, 385, 513, 641, 769, ... A great time-saver is provided by the observation that, for any number, the least divisor greater than 1 must be a prime number (a composite number, having lesser divisors of its own, cannot be the least). Consequently, in the investigation of F_5, we need not even bother with the composite 129. Since 257 is prime, it needs to be tried, but it does not divide. Again, 385 and 513 are composite, so they can be passed over. This brings us to the prime 641, which actually divides F_5. This procedure is based upon the work of a brilliant French number theorist, Edward Lucas, who published it in 1877. However, Euler knew almost as much a century and a half earlier. In 1747 one of his publications contained the result that every divisor of F_n is of the form $2^{n+1} \cdot k + 1$. (Lucas' improvement amounts only to the fact that k is always even.) Presumably he knew this in

1732 and used it to find the divisor 641. For F_5 we have $2^{n+1} \cdot k + 1 = 2^6 k + 1 = 64k + 1$, and for $k = 10$ we obtain the 641.

5. Pepin's Test. In the same year that Lucas determined the form of a divisor of F_n, T. Pepin published a remarkable test for settling whether F_n is prime or composite:

F_n divides $3^{(2^{(2^n-1)})} + 1$ if and only if F_n is a prime number.

Thus if one is able to show somewhow that F_n does not divide $3^{(2^{(2^n-1)})} + 1$, one may conclude that F_n is composite. It is in fact by showing that F_{1945} does not divide the number

$$m = 3^{(2^{(2^{1945}-1)})} + 1$$

that it was determined that F_{1945} is composite. This is a truly remarkable accomplishment. Do you have any idea how great F_{1945} is? It is so great that the number n which denotes how many digits it contains has, itself, over 580 digits: And F_{1945} is dwarfed by the almost inconceivably great number m into which it was shown not to divide. Nevertheless, Pepin's test was applied successfully, and in 1957 R. M. Robinson discovered that $5 \cdot 2^{1947} + 1$ is a divisor of F_{1945}. It is now known that F_n is composite for

$n = 5, 6, 7, 8, 9, 10, 11, 12, 13, 14, 15, 16, 18, 19, 21, 23, 25, 26, 27, 32, 36, 38, 39, 42, 55, 58, 63, 73, 77, 81, 117, 125, 144, 150, 207, 226, 228, 250, 267, 268, 284, 316, 452$, and 1945.

In 1905 J. C. Morehead used Pepin's test to show that F_7 is composite. In the same year A. E. Western, working independently, also proved this result. However, none of its divisors was known until 1971 when John Brillhart and Michael Morrison, using a computer at the University of California (Los Angeles), found the following two prime divisors:

$$\begin{aligned} F_7 = 2^{(2^7)} + 1 &= 2^{128} + 1 \\ &= 340282366920938463463374607431768211457 \\ &= 59649589127497217 \cdot 5704689200685129054721. \end{aligned}$$

In 1909 Morehead and Western, working together, also showed that

F_8 is composite. At present all of its divisors are still unknown. We conclude this essay with an outline of Morehead's attack on F_7. It is the same approach that was used with today's big computers to show that F_{1945} is composite.

Morehead needed to determine whether F_7 divides

$$3^{(2^{(2^7-1)})} + 1 = 3^{(2^{127})} + 1.$$

We observe that F_7 is a number of 39 digits. Now the least value of $3^{(2^n)}$ which is greater than F_7 is $3^{(2^7)}$, a number of 61 digits. Suppose that the remainder is r when F_7 is divided into $3^{(2^7)}$. Then r contains 39 or fewer digits. We may write

$$3^{(2^7)} \equiv r \pmod{F_7}.$$

Squaring gives $[3^{(2^7)}]^2 \equiv r^2 \pmod{F_7}$. Here r^2 certainly has no more than 78 digits. Let r_1 denote the remainder when r^2 is divided by F_7. Then we have

$$[3^{(2^7)}]^2 = 3^{(2^8)} \equiv r^2 \equiv r_1 \pmod{F_7},$$

where r_1 has no more than 39 digits. Squaring again gives

$$[3^{(2^8)}]^2 = 3^{(2^9)} \equiv r_1^2 \equiv r_2 \text{ (say) } \pmod{F_7},$$

where r_1^2 has no more than 78 digits and r_2 no more than 39 digits. Thus if we have the capacity to multiply two 39-digit numbers and to divide a 78-digit number by a 39-digit number, we can carry out this procedure and repeat it as often as we wish. Eventually we obtain

$$3^{(2^{127})} \equiv r_{120} \pmod{F_7},$$

where r_{120} has no more than 39 digits. Now, if and only if $r_{120} = F_7 - 1$ do we have

$$3^{(2^{127})} \equiv F_7 - 1 \pmod{F_7} \quad \text{and} \quad 3^{(2^{127})} + 1 \equiv F_7 \equiv 0 \pmod{F_7}.$$

Thus the thing to do is to compute r_{120} and see. Morehead undertook this Herculean task and found that r_{120} was not equal to $F_7 - 1$. (Can you imagine doing this on a 1905 desk calculator?) Consequently, F_7 does not divide $3^{(2^{127})} + 1$, implying that F_7 is composite.

Exercises

1. Prove that 2047 is a pseudoprime, and that 2821 is an absolute pseudoprime.

2. Prove that every two different Fermat numbers are relatively prime.

3. Show that every Fermat number except F_0 is of the form $12k + 5$.

4. For every natural number n, show that $2^{(2^n+1)} + 1$ is composite.

5. Show that every number in the following sequence is divisible by both 3 and 7:

$$a_1 = 2^{(2^2)} + 5, \quad a_2 = 2^{[2^{(2^2)}]} + 5, \ldots, \quad a_{n+1} = 2^{a_n-5} + 5.$$

6. If n is an odd pseudoprime exceeding 1, show that $2^n - 1$ is a (greater, odd) pseudoprime.

7. Prove that, if m is odd, then $2^m - 1$ and $2^n + 1$ are coprime for all natural numbers n.

8. Show that every natural number greater than 11 is the sum of two composite numbers.

References and Further Reading

1. W. Sierpinski, A Selection of Problems in the Theory of Numbers, Pergamon Press, 1964.

2. D. Shanks, Solved and Unsolved Problems in Number Theory, Spartan Books, Washington D.C., 1962.

3. W. Sierpinski, Elementary Theory of Numbers, Warszawa, 1964.

4. P. Erdös, On the converse of Fermat's theorem, Amer. Math. Monthly, 56 (1949) 623–4.

CHAPTER **2**

LOUIS PÓSA

1. The Story of Pósa. I would like to tell you something of the life and works of a remarkable young Hungarian named Louis Pósa (pronounced pō.sha), who was born in the late 1940's. When quite young he attracted the attention of the eminent Hungarian mathematician Paul Erdös (pronounced air.dish), who did much to help him develop. Erdös has recently written and spoken about some of the child prodigies he has known and I would like to tell you his story of Pósa.

In case you are not familiar with Erdös, let me first tell you a little about him. He is now about 60 years old and he has been for many years an internationally known mathematician. His three great interests are combinatorics, number theory, and geometry. He is a problem solver rather than a theory builder, although a fair number of his more than 500 mathematical papers exceed 100 pages in length. For decades he has travelled the world's universities, seldom visiting anywhere for more than a few months. He spent the Fall term of 1970 at the University of Waterloo, and during this visit he told us about the Hungarian prodigies. Except for a few minor changes and additions, the following is his story of Pósa.

Erdös' Story. "I will talk about Pósa who is now 22 years old and the author of about 8 papers. I met him before he was 12 years old. When I returned from the United States in the summer of 1959 I was told about a little boy whose mother was a mathematician and who knew quite a bit about high school mathematics. I was very interested and the next day I had lunch with him. While Pósa was eating his soup I asked him the following question: Prove that if you have $n+1$ positive integers less than or equal to $2n$, some pair of them are relatively prime. It is quite easy to see that the claim is not true of just n such numbers because no two of the n even numbers up to $2n$ are relatively prime. Actually I discovered this simple result some years ago but it took me about ten minutes to find the really simple proof. Pósa sat there eating his soup, and then after a half a minute or so he said "If you have $n + 1$ positive integers less than or equal to $2n$, some two of them will have to be consecutive and thus relatively prime." Needless to say, I was very much impressed, and I venture to class this on the same level as Gauss' summation of the positive integers up to 100 when he was just 7 years old."

At this point Erdös discussed a few problems in graph theory which he gave to Pósa. In order to avoid any misunderstandings, let us interject here a brief introduction to this material.

By a graph we do not mean anything connected with axes and coordinates. A graph consists of a collection of vertices (points) and a set of edges, each joining a pair of vertices. How the graph is pictured on paper is not essential. The edges may be drawn as straight lines or curves, and it is immaterial whether they are drawn so as to intersect or not. Points of intersection obtained by edges which cross do not count as vertices. Only the given vertices are the vertices of the graph. Also a graph need not contain every possible edge which could be drawn, that is, there are in general many different graphs with the same set of vertices.

A loop is an edge both of whose endpoints are the same vertex. Multiple edges occur when two or more edges join the same pair of vertices. In a general graph both loops and multiple edges are permitted. In a simple (or strict) graph neither loops nor multiple edges may occur. Throughout Erdös' story and the remainder of this essay the unmodified word graph is intended to mean simple graph. We continue now with the story.

"From that time onward I worked systematically with Pósa. I wrote to him of problems many times during my travels. While still 11 he proved the following theorem which I proposed to him: *A graph with $2n$ vertices and $n^2 + 1$ edges must contain a triangle.* Actually this is a special case of a well-known theorem of Turán, which he worked out in 1940 in a Hungarian labor camp. I also gave him the following problem: Consider an infinite series whose nth term is the fraction with numerator 1 and denominator the lowest common multiple of the integers $1, 2, \ldots, n$; prove that the sum is an irrational number. This is not very difficult, but it is certainly surprising that a 12 year old child could do it.

"When he was just 13, I explained to him Ramsey's theorem for the case $k = 2$: *Suppose you have a graph with an infinite number of vertices; then there is either an infinite set of vertices every two of which are joined by an edge, or there is an infinite set of vertices no two of which are joined by an edge.* (Incidentally, this theorem is the discovery of the late Frank Ramsey, a brother of the present Archbishop of Canterbury.) It took about 15 minutes for Pósa to understand it. Then he went home, thought about it all evening, and before going to sleep he had found a proof.

"By the time Pósa was about 14 years old you could talk to him as a grown-up mathematician. It is interesting to note that he had some difficulty with calculus. He never liked geometry, and he never wanted to bother with anything that did not really excite him. At anything that did interest him, however, he was extremely good. Our first joint paper was written when he was $14\frac{1}{2}$. Pósa wrote several significant papers by himself, some of which still have a great deal of effect. His best known paper, on Hamiltonian circuits, for which he received international acclaim, he wrote when he was only 15!

"The first theorem that he discovered and proved by himself which was new mathematics was the following: *A graph with n vertices ($n \geqq 4$) and $2n - 3$ edges must contain a circuit with a diagonal.* This result is the best possible, because for every n, one can construct a graph with n vertices and $2n - 4$ edges which fails to have a circuit with a diagonal.

"A problem which I had previously solved is the following: *A graph with n vertices ($n \geqq 6$) and $3n - 5$ edges must contain two circuits which have no vertices in common.* I told Pósa of the problem, and in a few days he had a very simple proof which was miles ahead of the complicated one I had come up with. A most remarkable thing for a child of 14.

"Pósa also found a beautiful proof that *every graph with n vertices and* $n + 4$ *edges contains two circuits which have no edges in common.* (This also holds for general graphs.)

"I would like to make a few conjectures why there are so many child prodigies in Hungary. First of all there has been for at least 80 years a mathematical periodical for high-school students. Then there are many mathematical competitions. The Eötvös-Kurshák competition goes back 75 years. After the first world war a new competition was begun for students just completing high school, and after the second war several new ones were started.

"A few years ago a different kind of competition was started. It is held on television. Bright high school students compete in doing questions in a given period of time. The questions are usually very ingenious and the solutions are judged by a panel of leading mathematicians such as Alexits, Turán, and Hajos. It seems many people watch these competitions with great interest even though they do not understand the problems.

"In Hungary a few years ago a special high school, the Michael Fazekas High School, was opened in Budapest for children who are gifted in mathematics. The school started just when Pósa was due to go to high school. He liked the school very much, so much so, in fact, that he refused to leave it for entrance into university two years early. Soon after attending Fazekas High School, Pósa was telling me of other boys in his class who he thought were better at elementary mathematics than he was. Among these boys was the now prominent Lovász."

It seems appropriate to complement the story of Pósa with a sample of his work. Accordingly, let us work through his beautiful proof of Dirac's theorem on Hamiltonian circuits.

2. Dirac's Theorem. Graphs have many interesting properties. In 1857 the Irish genius William Rowan Hamilton invented a game of travelling around the edges of a graph from vertex to vertex. Given a particular graph to begin with, the object of the game was to find a path in the graph which passes through each vertex exactly once.

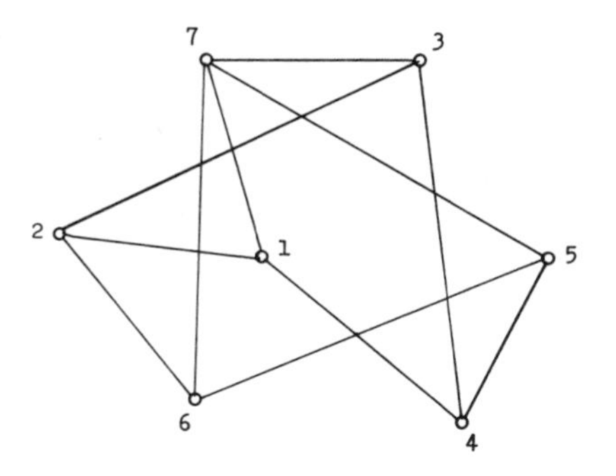

Of course, in some graphs such a path does not exist. If, in addition to finding such a path, one can arrange to make the last vertex the same as the first, he obtains a Hamiltonian *circuit,* not just a Hamiltonian *path.* While a Hamiltonian circuit always provides a Hamiltonian path, upon the deletion of any edge, a Hamiltonian path may not lead to a Hamiltonian circuit (it depends on whether or not the first and last vertices of the path happen to be joined by an edge in the graph).

There are some necessary and some sufficient conditions known for the existence of a Hamiltonian circuit in a graph. For example, it is necessary to have as many edges as vertices, and it is sufficient to have all possible edges actually in the graph. However, even in 1972 there are no known necessary and sufficient conditions.

In 1952 the European mathematician G. A. Dirac came up with the following simple sufficient condition. (The "valence" of a vertex is simply the number of edges that occur at the vertex.)

DIRAC'S THEOREM. *A graph with n vertices* ($n \geqq 3$) *in which each vertex has valence at least* $n/2$ *has a Hamiltonian circuit.*

In 1962 Pósa produced the following proof. He proceeds indirectly. Suppose the given graph G which has n vertices and every valence at least $n/2$ is non-Hamiltonian, that is, does not possess a Hamiltonian circuit. We now proceed to a contradiction.

Firstly we embed G in a saturated non-Hamiltonian graph G' as follows:

Clearly not all possible edges occur in G or else it would be Hamiltonian already. Consider, then, the insertion of an edge which is missing from G. If this insertion does not complete a Hamiltonian circuit. then leave the edge in the graph. It if does complete a Hamiltonian circuit, take the edge back out. Go around G doing this at each missing edge. The order in which one tries the missing edges is unimportant. At the end of this tour, the resulting graph G' will still be non-Hamiltonian since no Hamiltonian circuit is ever allowed to remain intact, and yet it will be saturated in the sense that the addition now of any missing edge will have to complete a Hamiltonian circuit (otherwise we would have left the edge in the graph when it was tried during our rounds).

Adding edges certainly does not diminish any valence. Hence every vertex of G' has valence at least $n/2$.

Since G' is non-Hamiltonian, it cannot contain all possible edges. Suppose the edge between vertices v_1 and v_n is missing. Putting in this edge would complete a Hamiltonian circuit because G' is saturated. Thus, even without the edge v_1v_n, the graph G' must contain a Hamiltonian *path* from v_1 to v_n (inserting the edge v_1v_n merely completes this path into a circuit). Let us denote the order of the n vertices in this path by $v_1, v_2, v_3, \ldots, v_{n-1}, v_n$.

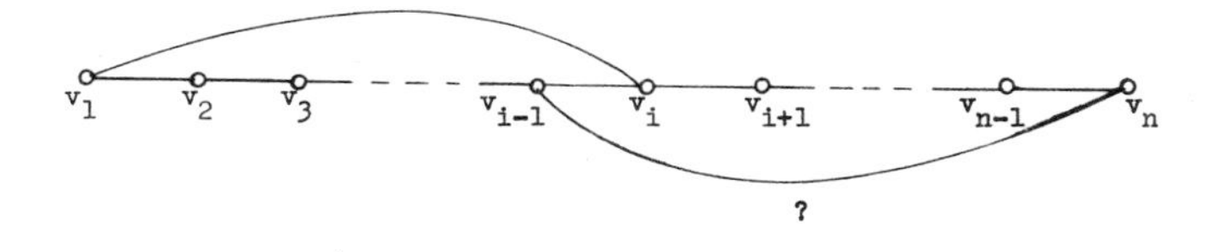

Now if the vertex v_1 happens to be joined to the vertex v_i, we ask whether the vertex v_n could possibly be joined to v_{i-1}? The answer is "no." For, if v_n were joined to v_{i-1}, then G' would contain the

Hamiltonian circuit

$$v_1 v_i v_{i+1} \cdots v_n v_{i-1} v_{i-2} \cdots v_2 v_1.$$

This is impossible since G' is non-Hamiltonian.

While v_1 is not joined to v_n, we do know that v_1 has valence at least $n/2$. Thus v_1 is joined to at least $n/2$ vertices v_i where i runs from 2 to $n - 1$. Consequently, there are at least $n/2$ vertices v_{i-1}, where $i - 1$ ranges from 1 to $n - 2$, to which v_n cannot possibly be joined. Since loops are not permitted, v_n cannot be joined to itself, either. Altogether, then, there are more than $n/2$ vertices to which v_n cannot be joined. This leaves fewer than $n/2$ vertices to which v_n can be joined, contradicting the given fact that its valence is at least $n/2$. (QED)

We conclude this chapter with the statement of four recent theorems in the field of Hamiltonian circuits. The final one is the work of V. Chvátal, an energetic young Czech mathematician, who produced it during his stay at the University of Waterloo, where shortly before he had earned his Doctorate Degree.

DIRAC'S THEOREM (1952). *A graph with n vertices ($n \geqq 3$) in which each vertex has valence at least $n/2$ has a Hamiltonian circuit.*

ORE'S THEOREM (1960). *If $n \geqq 3$ and, for every pair of vertices that are not joined by an edge, the sum of the valences is at least n, then the graph is Hamiltonian.*

PÓSA'S THEOREM (1962). *Let G be a simple graph with n vertices. If for every k in $1 \leqq k < (n - 1)/2$, the number of vertices of valence not exceeding k is less than k, and if for n odd the number of vertices with valence not exceeding $(n - 1)/2$ does not exceed $(n - 1)/2$, then G contains a Hamiltonian circuit.*

CHVÁTAL'S THEOREM (1970). *Let the graph G have vertices with valences $d_1, d_2, \ldots, d_n$, written in non-decreasing order. If for every $i < n/2$ we have either $d_i \geqq i + 1$ or $d_{n-i} \geqq n-i$, then the graph is Hamiltonian.*

Exercises

1. Find a Hamiltonian circuit in each of the following graphs:

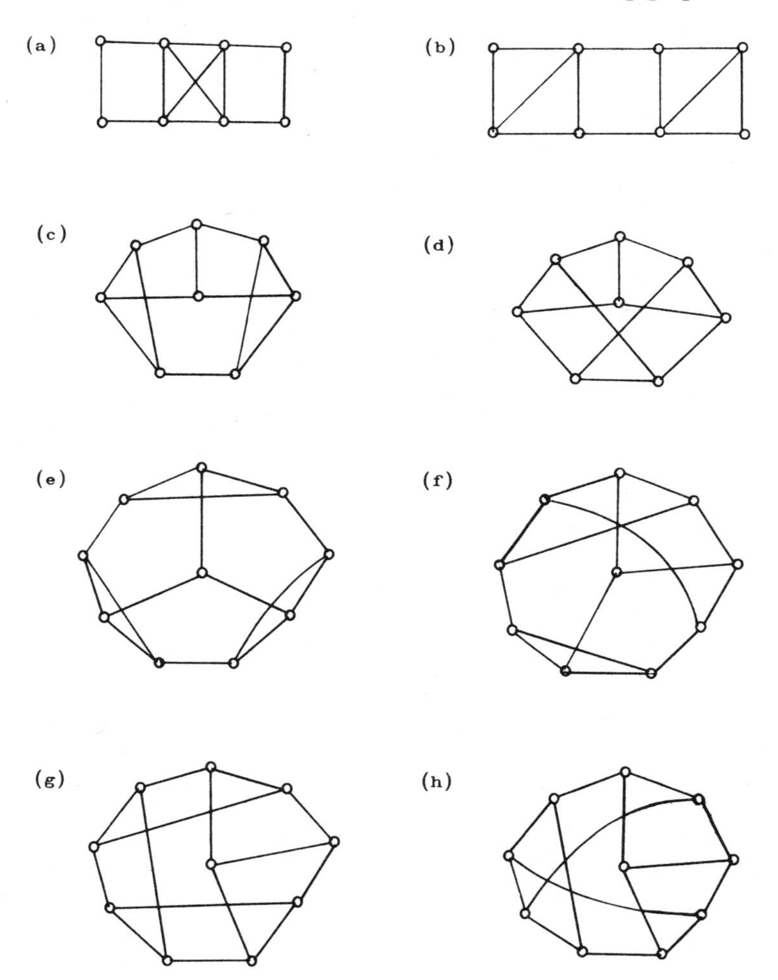

2. Prove that there is no Hamiltonian circuit in each of the fol-

lowing graphs:

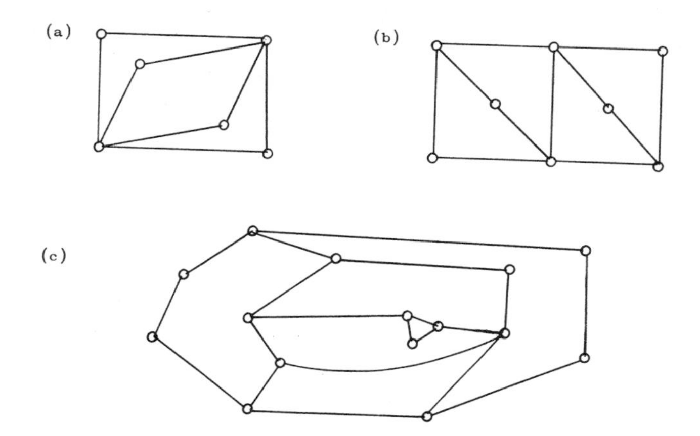

3. Prove there is no Hamiltonian *path* in either of the graphs

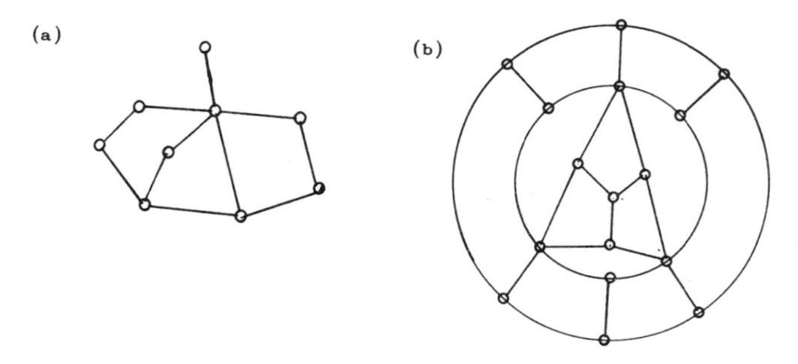

4. Which of the following graphs have Hamiltonian circuits and which have only Hamiltonian paths? (See figures on page 19.)

5. Six points in general position are given in three-dimensional space (that is, no 3 are collinear and no 4 are coplanar). The $\binom{6}{2} =$ 15 segments joining them in pairs are colored individually either red or blue at random. Prove that some triangle has all its sides the same color.

6. A set of moves in chess which takes a knight successively

through all 64 squares is called a knight's tour. If the knight can go from the last square to the first one in one move, and thus go all around again, the tour is called re-entrant. A re-entrant knight's tour corresponds to a Hamiltonian circuit in a graph which has a vertex for each square and an edge joining the vertices represent-

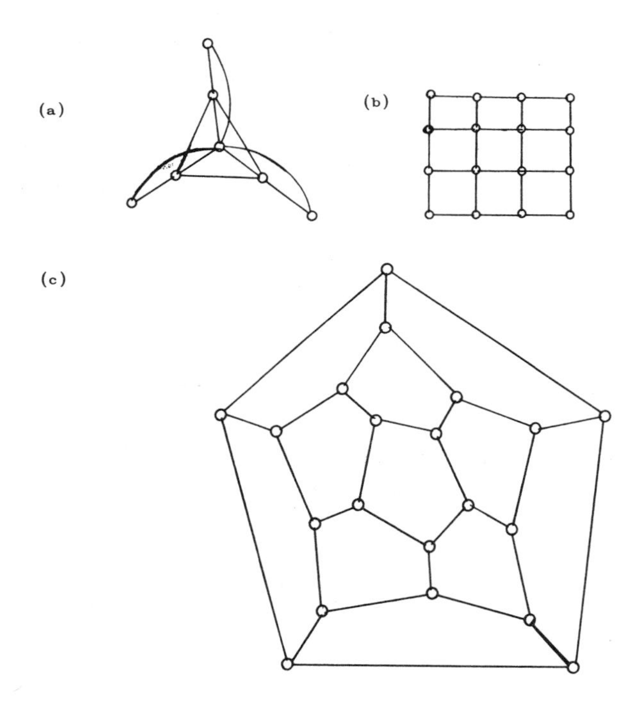

ing squares X and Y if and only if a knight can go from X to Y in one move. Show that there is no re-entrant knight's tour on any chessboard which has dimensions 4 by n, n a natural number.

7. Model a proof of Ore's Theorem (above) after Pósa's proof of Dirac's Theorem.

8. Try your hand at proving some of the theorems Pósa worked through as a teenager. For example, prove that a graph with n vertices and $n+4$ edges contains two circuits with no edge in com-

mon. The proof of this is contained in the joint paper of Erdös and Pósa "On the Maximum Number of Disjoint Circuits of a Graph", Publications Mathematicae, Debrecen, 1962, 3–12, especially page 9. This journal may be listed under Kossuth Lajos Tudomanyegyetem, Mathematika Intezet.

9. Prove Ramsey's Theorem. If you have trouble, consult the following plan. Consider an infinite sequence of vertices, 1, 2, 3, ...

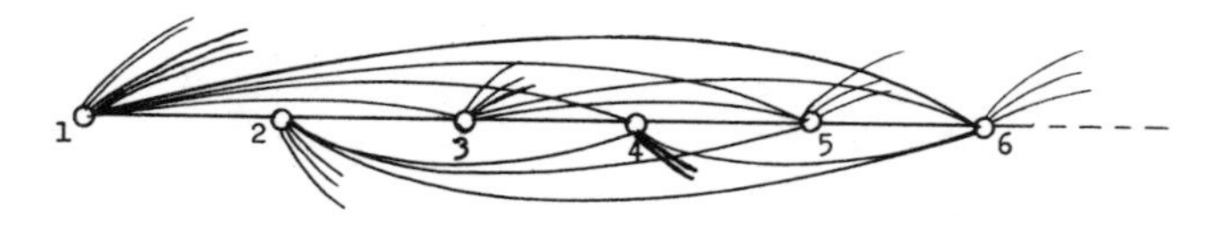

Suppose each pair is joined by an edge. The edge joining vertices n and $n + p$, $p > 0$, is considered to begin at n and end at $n + p$. Now all edges are colored either red or blue so that, for every vertex n, all the edges which begin at n have the same color. If all are red, then n is called a red vertex, etc. Which vertices are red and which blue is immaterial.

(a) Show in such a sequence there exists either an infinite set of vertices every two of which are joined by a red edge, or there exists an infinite set of vertices every two of which are joined by a blue edge. (Hint: Consider the set of red vertices and the set of blue vertices; at least one of these sets must be infinite.)

(b) Now complete the proof of Ramsey's theorem by showing that in every graph which has an infinite number of vertices one can determine a sequence of vertices as described above.
(Hint: Color all edges of the graph red. Then insert all the missing edges and color them blue. Take any vertex at all as vertex 1 of the sequence. Separate into classes R and B the vertices to which 1 is joined, respectively, by red and blue edges. At least one of R and B must contain an infinity of vertices. If both do, choose either. For definiteness, suppose R is infinite. Then take any vertex of R as vertex 2 of the sequence.

Repeat the separation process with respect to vertex 2 and its infinite set R (forget all about set B). Choose any vertex of an infinite set thus obtained as vertex 3 of the sequence, and so on. From this sequence Ramsey's theorem is easily deduced by part (a). At the end rub out all the blue edges.)

10. Let $A_0, A_1, A_2, \ldots, A_{2n-1}$ denote, in cyclic order, the vertices of a regular $2n$-gon. Let all the sides and diagonals be drawn to give graph G. Prove that every Hamiltonian circuit of G must contain two edges which are parallel lines in the diagram.

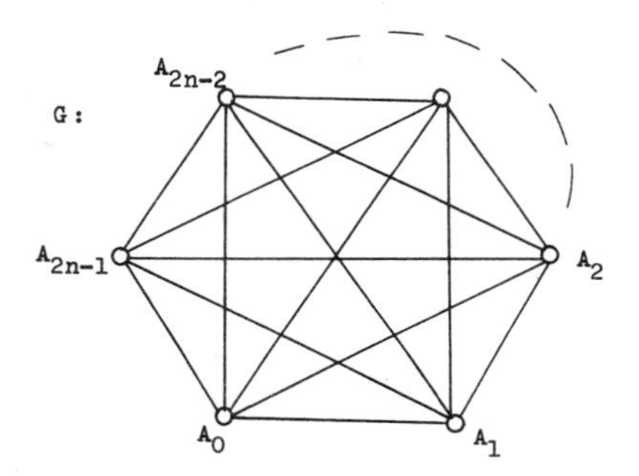

References and Further Reading

1. Behzad and Chartrand, Introduction to the Theory of Graphs, Allyn and Bacon, Boston, 1972.

2. F. Harary, Graph Theory, Addison-Wesley, Reading, 1969.

CHAPTER **3**

EQUILATERAL TRIANGLES

In this essay I would like to tell you about three outstanding problems involving equilateral triangles. We begin by reviewing a few easy properties of equilateral triangles. Thus, when we refer to them later, we will not need to interrupt our story in order to discuss them.

1. Simple Properties. Every triangle has a circumcenter, an incenter, a centroid, etc., so that in general one cannot speak of "the" center of a triangle. In the case of an equilateral triangle, however, all these various center-points coincide, giving an unambiguous meaning to the term.

Letting s and a denote, respectively, the lengths of the side and altitude of an equilateral triangle, we see by the Pythagorean theorem that $s^2 = a^2 + s^2/4$, which gives

$$a = \frac{\sqrt{3}}{2} s \quad \text{and} \quad s = \frac{2}{\sqrt{3}} a.$$

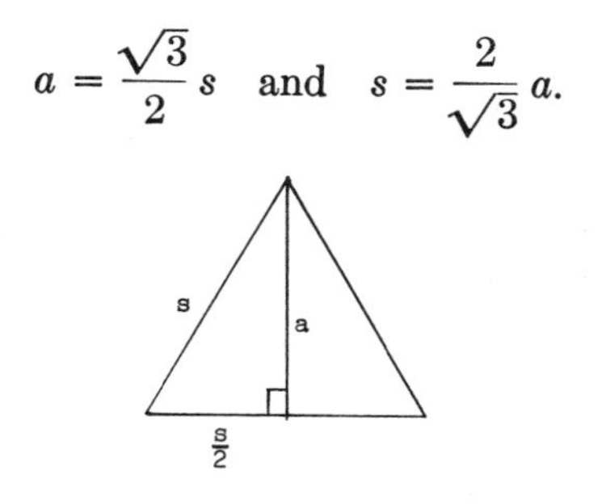

We observe that an isosceles triangle which possesses a $60°$ angle between its equal sides is in fact equilateral. We note also that the angle subtended by a side of an equilateral triangle at a point on the circumcircle is $60°$ or $120°$, depending on whether the point occurs on the major or minor arc cut off by the side.

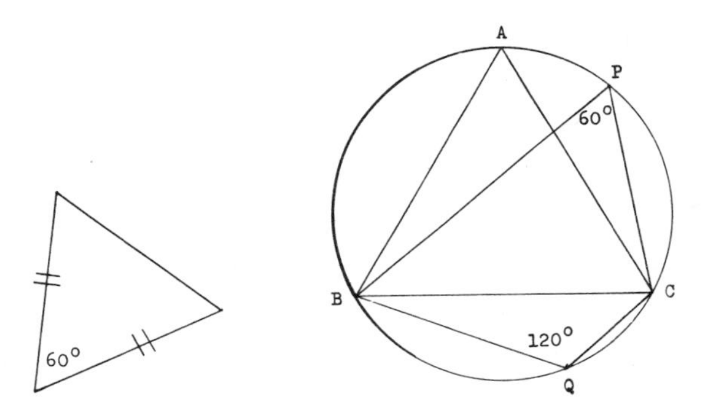

Finally we establish a very simple result known as Viviani's theorem. (Viviani: Italian, 1622–1703.)

VIVIANI'S THEOREM. *For a point P inside an equilateral triangle ABC the sum of the perpendiculars a, b, c from P to the sides is equal to the altitude h.*

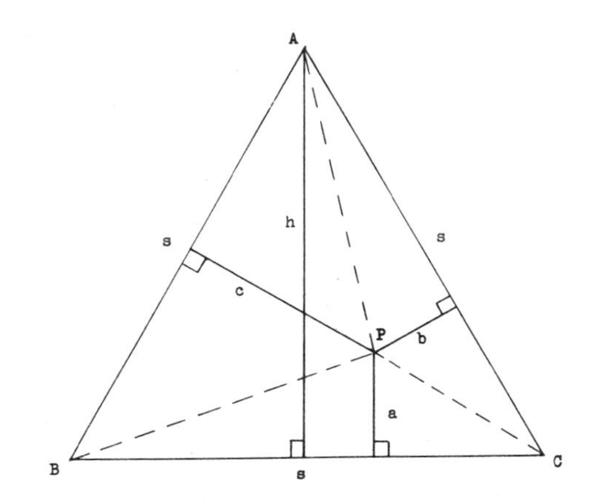

Proof. We have in terms of areas

$$\Delta ABC = \Delta PBC + \Delta PCA + \Delta PAB$$

$$\tfrac{1}{2}sh = \tfrac{1}{2}sa + \tfrac{1}{2}sb + \tfrac{1}{2}sc$$

$$h = a + b + c.$$

2. Fermat's Problem to Torricelli. The famous French number theorist Pierre de Fermat (1601–1665) gave the following problem to Evangelista Torricelli (1608–1647), the well-known student of Galileo and discoverer of the barometer:

Determine a point P in a given triangle ABC such that the sum $PA + PB + PC$ is a minimum.

Torricelli solved the problem several ways. We consider his simplest and most exciting solution which is based upon Viviani's theorem and which was rediscovered almost 300 years later by the Hungarian mathematician Frederick Riesz.

It turns out that if the given triangle contains an angle of 120° or more the required point P is the vertex of this obtuse angle. We shall not consider this case in favor of Torricelli's most elegant solution in the case of a triangle with no angle as great as 120°.

As one might expect, the point which solves the problem has, down through the years, come to be known as the Fermat point of the triangle. Various ways of specifying its location have been discovered. For example, at the Fermat point each side of the triangle subtends an angle of 120°. Let us take this as its definition, and proceed to show that it solves our problem. On this basis, the problem of actually finding the position of the Fermat point for a given triangle is easily solved. All one need do is draw equilateral triangles outwardly on two sides and note where their circumcircles meet.

Let us call this Fermat point P. Torricelli's first step is to draw around the given triangle ABC a triangle XYZ by drawing perpendiculars to PA, PB, and PC. In quadrilateral $ZAPB$, then, the angle Z is $360° - 90° - 90° - 120° = 60°$. Similarly for the angles at Y and X, making ΔXYZ equilateral. Thus by Viviani's theorem $PA + PB + PC = h$, the altitude of ΔXYZ.

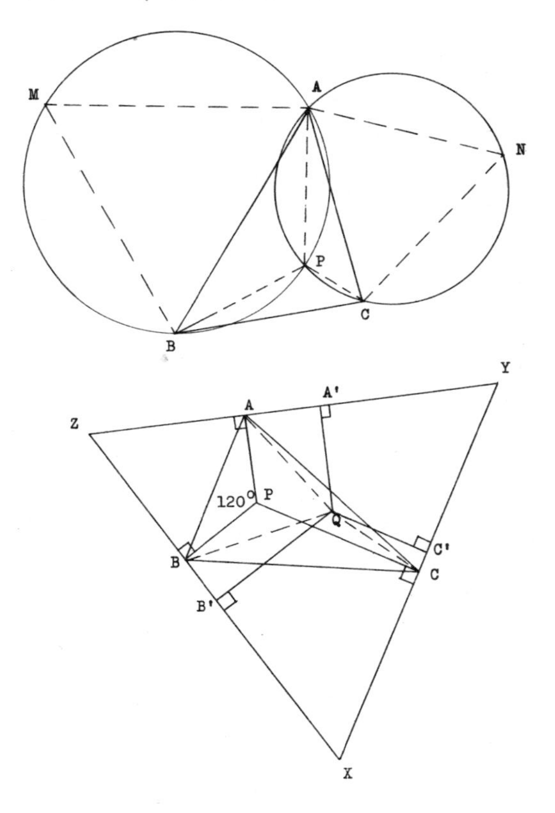

For any other point Q in ΔABC we also have that the sum of the perpendiculars QA', QB', and QC' to the sides of ΔXYZ is h. But in general the hypotenuse QA of right triangle QAA' exceeds leg QA'. Similarly QB and QC, respectively, exceed legs QB' and QC'. Thus the sum

$$QA + QB + QC > QA' + QB' + QC' = h = PA + PB + PC,$$

showing P to be the solution to our problem.

We observe that at most one of the right triangles can collapse and yield the equality of a hypotenuse and leg (e.g., when Q lies on PA). Thus at most one of QA', QB', QC' can actually be as great as the corresponding QA, QB, QC, giving universal validity to the inequality

$$QA + QB + QC > QA' + QB' + QC'.$$

Another ingenious solution, involving equilateral triangles, for the case in which each angle of the triangle is less than 120 degrees, was published by J. E. Hofmann in 1929. The method was not new at the time and credit for it must be shared with the Hungarian mathematician Tibor Gallai and others who discovered it independently. The first step is to choose any point P in the given triangle ABC. Then ΔPAB is rotated outwardly through $60°$ about vertex B to a position $C'P'B$. This makes triangles $C'AB$ and $P'PB$ isosceles with $60°$ angles between their equal sides. Thus both these triangles are equilateral. That is to say, no matter where the point P is taken in ΔABC, the rotation always carries A to the same place C', which is just the third vertex of an equilateral triangle drawn outwardly on AB.

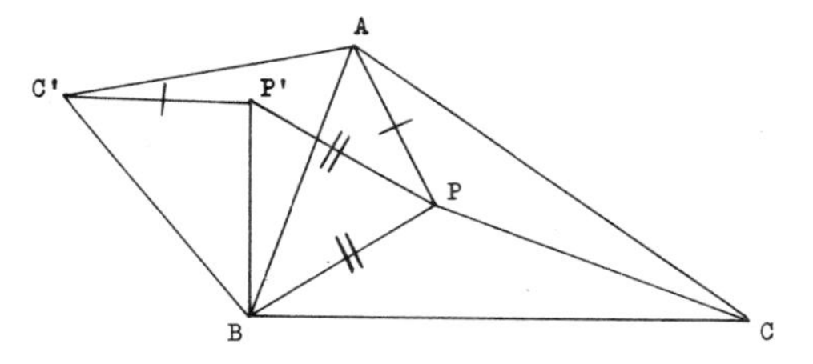

Since $\Delta P'PB$ is equilateral, we have $PB = PP'$. The rotation carries AP into $C'P'$, making $AP = C'P'$. Consequently

$$PA + PB + PC = C'P' + P'P + PC = \text{path } C'P'PC.$$

For all choices of P, this path begins at the same point C' and ends at the same point C. If, for some choice of P, the path is straight, the problem will be solved.

At this point in our investigation we are not sure that a straight path will ever occur. In the event that $C'P'PC$ were to be straight, the point P would have to occur on the line $C'C$ and the angle BPC' would have to be 60°. Let us, then, join C' to C and note the point P where it crosses the circumcircle of the equilateral triangle $C'AB$. In this circle the angle BPC' would be 60°. Now the rotation always carries the point P to a point P' such that $\angle BPP' = 60°$. Accordingly, for P the intersection of $C'C$ and the circumcircle, the image P' occurs on $C'P$ and $C'P'PC$ is indeed straight. This choice of P, then, solves the problem.

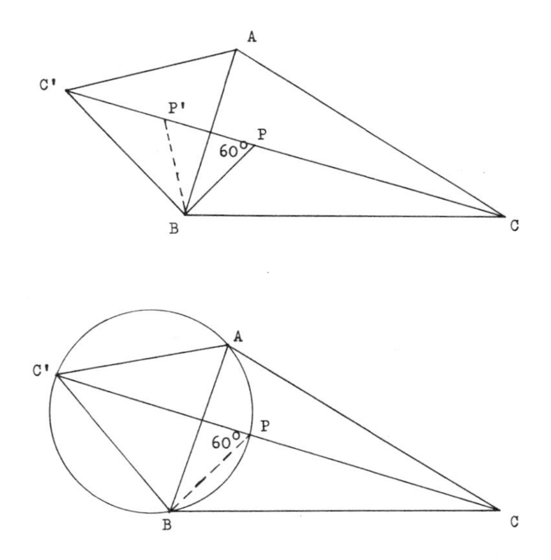

Torricelli's solution clearly shows that the solution to our problem is unique. Thus Hofmann's solution gives us another specification of the Fermat point of a triangle. Accordingly, it must lie on the line $C'C$ joining the third vertex C' of an equilateral triangle drawn outwardly on the side AB and the opposite vertex C. However, there is nothing special about C and C'. We could equally well have rotated ΔPBC outwardly through 60° about vertex C, or ΔPAC about A, to yield the same result concerning vertices A

and B. With regard to the three segments which join the vertices A, B, C to the third vertices A', B', C' of equilateral triangles drawn outwardly on the opposite sides, we obtain the corollary:

The segments AA', BB', CC' are all the same length (equal to the minimal sum of PA, PB, PC), are concurrent at the Fermat point P, and meet there at 60° angles.

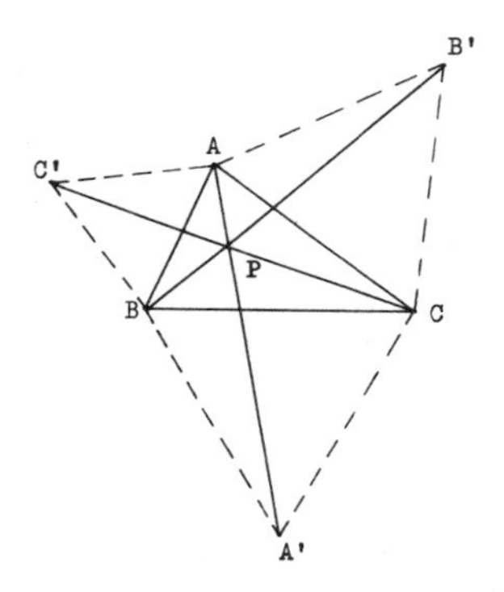

In solving problems on maxima and minima we are generally faced with three questions (i) Does a solution exist? (ii) Is the solution unique? (iii) What properties characterize the solution(s)? Torricelli's work shows that the point P at which each side of the given triangle subtends 120 degrees gives a smaller value for $PA + PB + PC$ than any other point of the triangle. Thus he disposes of all three aspects of the problem at one stroke. This is certainly very nice. But it is unthinkable that Torricelli's first impulse was to try out this point P. What ever made him think of this point P? We shall probably never know, but in the nineteenth century the great Swiss geometer Jacob Steiner described a very beautiful sequence of ideas which led him to think of it. His thought proceeded as follows.

First he wondered if a vertex could solve the problem. For P at a vertex, one of the terms in $PA + PB + PC$ vanishes, and the other two are sides of the triangle. In fact, the value of the sum is just the sum of the two sides which meet at the vertex under investigation. Consequently, the best vertex is the one at which the two smallest sides meet; the biggest side should be avoided. Thus, if the solution is a vertex, it is the vertex of the largest angle, which is opposite the longest side.

Next he tried to discover any properties of a point P, which is not a vertex, that solves the problem. Let us denote the distances

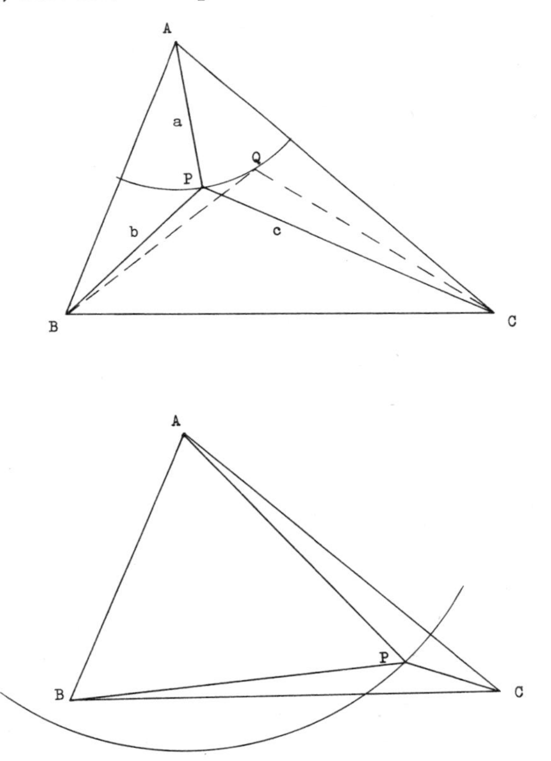

PA, PB, PC by a, b, c, respectively, and their sum by S. Steiner employs a very useful principle which applies to many kinds of extremal problems. It is often necessary that a solution to an extremal problem provide the same extreme value for a subset of the variables which remain after other variables in a solution are kept fixed. For example, it is necessary to minimize every leg of a journey in order to minimize the entire journey. Clearly any point P which minimizes S provides a particular set of values for a, b, and c. If a is kept fixed at this particular value, then it must be that the solution P provides the smallest possible sum for $b + c$ that can be obtained for points Q with $QA = a$. That is to say, of all the points Q on the circle with center A and radius a, no point Q can provide a smaller sum than P for the quantity $QB + QC$ (which is $b + c$ for $Q = P$). This leads to an interesting analysis.

We begin by showing that the vertices B and C must lie outside this circle with center A and radius a. For the vertex B, the value of S is the sum of the sides $AB + BC$. Now $PB + PC$ is greater than BC, and if vertex B were to occur inside the circle, or even on the circumference, the radius PA would be at least as big as the other side AB. As a result, the vertex B would provide a smaller value of S, namely $AB + BC$, than the point P, for which $S = PA + (PB + PC)$. Consequently, the vertex B, and likewise C, must fall outside the circle.

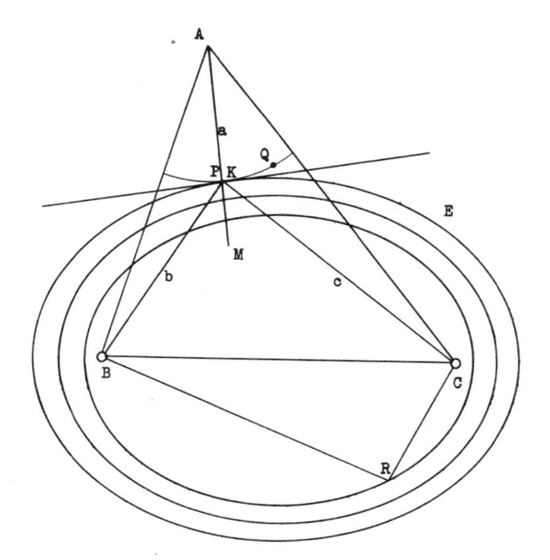

Consider now a small ellipse with foci at B and C. At each point R of the ellipse we obtain the same sum for the focal radii $RB +$

RC. We can imagine this ellipse to have been drawn with a pencil which keeps taut a string fastened at B and C. If we permit the string to get a little bigger, a bigger ellipse is obtained which completely surrounds the former ellipse. Using longer and longer strings, we obtain ellipses of all sizes. Let E denote the ellipse of this family which just touches the circle under investigation. Let the point of contact be K. Since every other point of the circle is

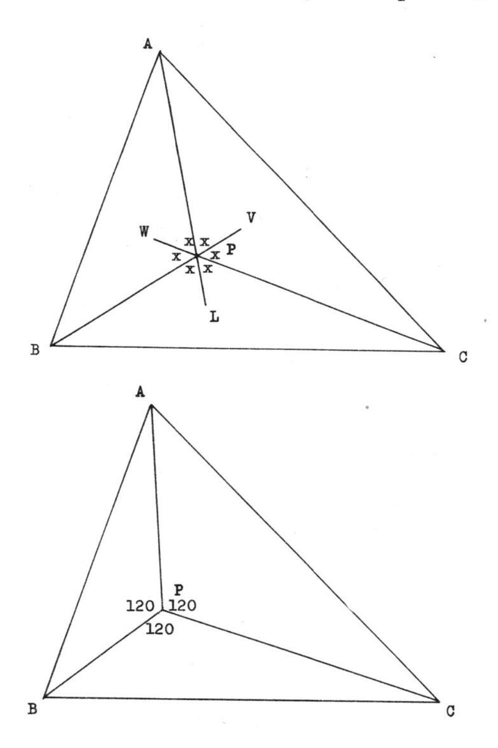

outside E, it would take a longer string to put an ellipse through any of them. But the length of the string is the sum of the focal radii. Thus the point of contact K provides the smallest sum $QB + QC$ for points Q on the circle. Since K is the only point of contact of E and the circle, it is the only point of the circle to minimize

$QB + QC$. Since P also provides this minimum, it must be that $P = K$.

Thus the circle and ellipse have a common tangent at P. The radius AP is perpendicular to this tangent, and thus it gives the normal to the ellipse at the point P. However, the well-known reflector property of the ellipse tells us that a ray through one focus reflects through the other focus, implying that the angle of incidence, BPM, equals the angle of reflection, MPC. That is to say, at a solution point P, the line AP extended bisects the angle BPC. But there is nothing special about AP. The same must be true of BP and CP, that is, their extensions must bisect, respectively, the angles APC and APB.

Let us denote the equal angles BPL and LPC by x. The angle $APV = x$ because it is vertically opposite angle BPL. The bisector PV, then, makes angle VPC also equal to x. Similarly each half of angle APB is x. This gives $6x = 360$ degrees, and $x = 60$ degrees. Consequently, each side of the triangle subtends an angle of 120 degrees at the solution point P (see p. 31). Steiner carries on to other conclusions but we shall leave him at this point.

In his little book *Circles*, Daniel Pedoe offers a solution to our problem which is based upon an extension of Ptolemy's theorem. We conclude our story of this problem with a neat solution taken from the realm of mechanics. As Archimedes discovered over 2000 years ago, the consideration of apparently remote notions like mass and energy can lead to handsome rewards in pure mathematics. Again we restrict ourselves to the case in which no angle of the given triangle is as large as 120 degrees.

We suppose that the given triangle ABC lies in a horizontal plane and that it is outfitted with a smooth pulley at each vertex. From a point P in the triangle strings are passed over the pulleys and a mass m is suspended at the end of each string. The system is released and allowed to settle under the influence of gravity. The position taken by the point P, then, is the required Fermat point of ΔABC. We argue as follows.

In the position of equilibrium, let the masses m occur at heights a, b, c above a horizontal plane π, and let the centroid G of the system be a distance r above π. The total potential energy of the system relative to π is given by either side of the equation

$$ma + mb + mc = (3m) \cdot r.$$

Thus we have $r = \frac{1}{3}(a + b + c)$.

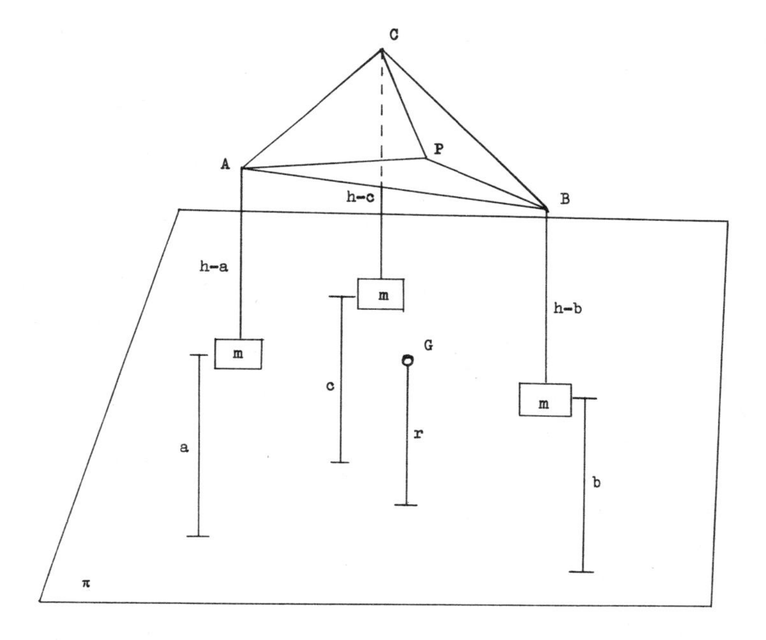

However, the position of equilibrium represents that of minimal potential energy. Consequently, at this position, G is lowest, that is, closest to π, and $a + b + c$ is minimal. Let h denote the distance of ΔABC above π. Then the vertical parts of the strings have lengths $h - a$, $h - b$, $h - c$, with sum $3h - (a + b + c)$. If the entire length of string is t, then the length of horizontal string on ΔABC is given by

$$s = t - [3h - (a + b + c)] = (t - 3h) + (a + b + c).$$

Since t and h are constants and $a + b + c$ is minimal at the position of equilibrium, s is also minimal there.

To convert this into a nice geometric specification for the Fermat point P, we note that P is held in equilibrium by three equal tensions f in the strings (the same mass m is on each string).

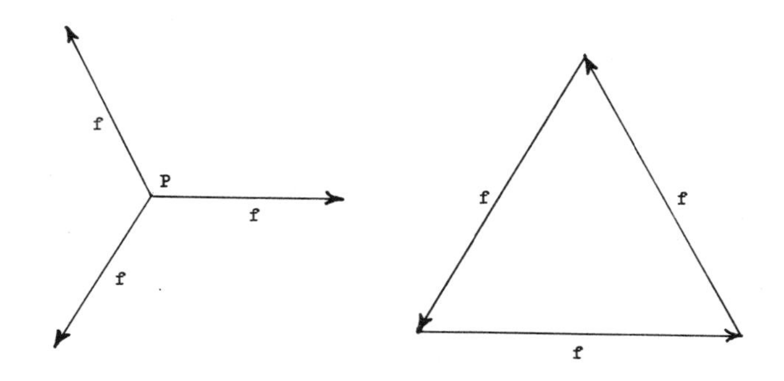

Consequently, the three force vectors f representing these tensions must close an equilateral triangle when added vectorially. This implies that the angles between the strings is 120 degrees, giving our original specification of the Fermat point.

We note that this mechanical procedure provides the "minicenter" of a set of n points in the plane which determine a convex n-gon. However, a nice geometric description of its location does not emerge in general. There is only one way to make a plane polygon with three equal sides, namely an equilateral triangle. For n greater than 3, however, there are generally many ways to close an equal-sided n-gon. As a result, the angles between the tension vectors are not determined uniquely.

3. Napoleon's Theorem. Historically the following theorem is known as Napoleon's theorem, although it is very doubtful that Napoleon was well enough versed in geometry to have discovered and proved it himself.

Let $C'AB$, $A'BC$, $B'CA$ denote equilateral triangles drawn outwardly on the sides of given triangle ABC; then the centers X, Y, Z of the equilateral triangles themselves form an equilateral triangle.

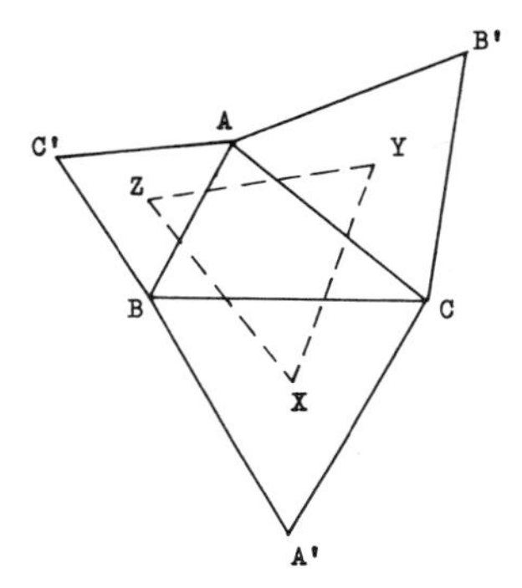

I believe the following is a new proof of this theorem:

We begin by drawing the circumcircles of the equilateral triangles. Only the major arcs of these circles are of use to us, and we do not need the equilateral triangles at all. Let PQ denote the segment intercepted by the arcs on a line through A. Then PB and QC make 60° angles at P and Q. Consequently, PB and QC extended meet at a point R at which another 60° angle must occur. At this point R, then, the side BC subtends a 60° angle, putting R on the third arc. That is to say, for every point P on the arc AB

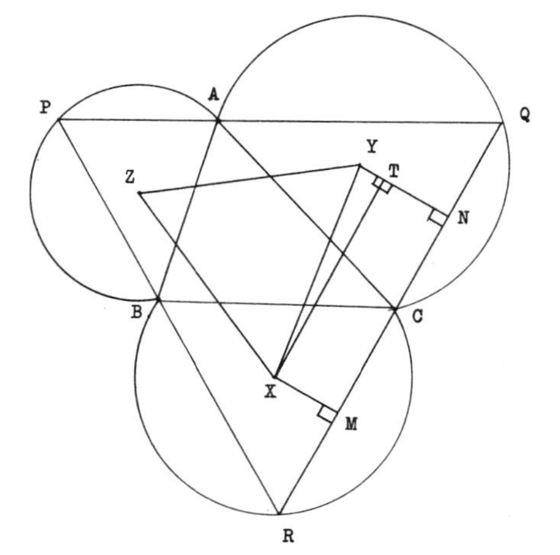

there is an equilateral triangle PQR which circumscribes the given triangle ABC, and has a vertex on each of the arcs. Our theorem follows immediately from a simple consideration of the largest member of this family of equilateral triangles.

In order to get a line on the size of a circumscribing triangle, construct the perpendiculars to QR from centers X and Y to give the midpoints M and N of chords RC and CQ. Let XT complete rectangle $XMNT$. Clearly QR is twice MN, which is the same as twice XT. However XT is a leg of right triangle XYT and is therefore not greater than hypotenuse XY. When QR is parallel to XY, the right triangle collapses, making $XT = XY$, and QR attains its maximum length of twice XY.

Similarly the maximum of side PQ is twice YZ and that of PR is twice XZ. Since PQR is always equilateral, its sides are always equal and therefore they have the same maxima. This means $2XY = 2YZ = 2ZX$, implying ΔXYZ is equilateral.

4. A Covering Problem. Our final problem concerns a family of sets of points in the plane, the so-called sets of "diameter 1." If d is the smallest number about which it can be said that "no two points of the set are farther apart than d," then d is the diameter of the set. The set of points determined by the circumference and interior of a circle of radius 2 has diameter 4. If we drop the circumference from this set, the remaining interior still has diameter 4. Even though no two points of the set are as far apart as 4, there is for every number s which is less than 4 some pair of points which are farther apart than s. Formally, the diameter is the least upper bound of the set of distances between pairs of points of the set. This is identical with the maximum distance when a maximum exists.

In this problem we are concerned with the entire family of sets of diameter 1. This includes an infinite variety of shapes and sizes, of continuous regions and isolated points. It taxes the mind even to imagine a typical member of this family. Nevertheless it is quite easy to prove the following remarkable result concerning these sets:

Every plane set of diameter 1 *can be completely covered with an equilateral triangle of side* $\sqrt{3}$.

Since the ability to cover a set of points depends on the size and shape of the set, we are quite surprised to be able to prove such a result at all. The diameter of the set seems to carry a minimum of information concerning these aspects of the set. However, when we realize that the size of the incircle of an equilateral triangle of side $\sqrt{3}$ is itself of diameter 1, we feel the immediate relief suggested by the procedure of first covering the given set with a circle of diameter 1 and then covering the circle with the equilateral triangle. Unfortunately we are soon brought up short with the surprising fact that it is not always possible to cover a set of diameter 1 with a disk of diameter 1. We see this by noting that the set of points determined by an equilateral triangle of side 1 (a set of diameter 1) cannot be covered with a circle of diameter 1; its circumcircle has diameter $2/\sqrt{3}$, which is approximately 1.15. This shock informs us that we really do have a nontrivial problem before us. However, it does have a beautiful and simple solution.

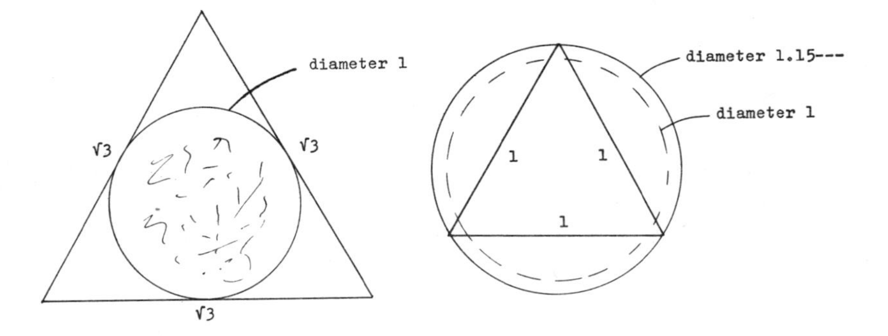

Clearly the given set S can be enclosed in many strips in the plane. In each direction there is a narrowest strip which contains S. Let the bounding edges of such a minimal strip be m_1 and m_2.

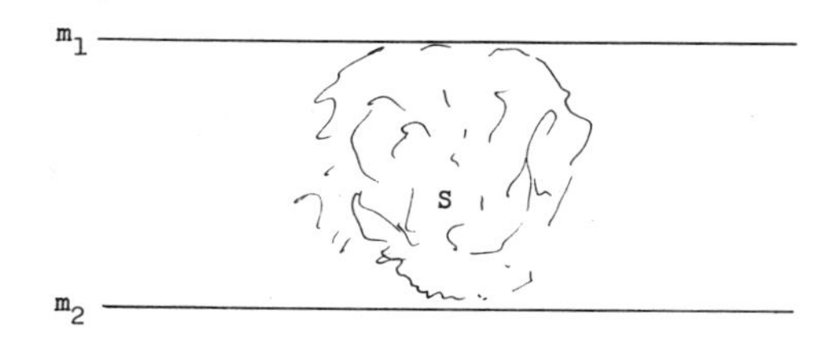

Now two cases occur: there may or may not be points of S on m_1 and/or m_2. If S does not contain a boundary (like the interior of a circle) then the lines m will not actually make contact with S. In any event we can say with certainty that even the slightest narrowing of this minimal strip will cause S to overflow the strip. Because of this we are able to deduce easily that the distance across a minimal strip m_1m_2 is at most 1 unit.

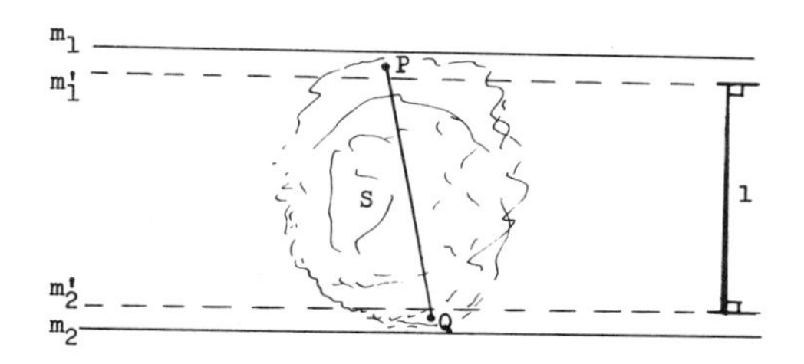

If the strip m_1m_2 were to be wider than 1 unit, it could be narrowed to exactly 1 unit by moving m_1 and m_2 together, moving each the same amount. This would cause S to overflow the narrowed strip on both sides. Consequently there must exist some pair of points P and Q of the set S which are farther apart than the diameter 1, which is the width of the narrow strip. But this is impossible.

Let aa', bb', cc' denote any three minimal strips containing S which are inclined to each other at 60° angles. These strips provide two equilateral triangles abc and $a'b'c'$ which enclose the given set S. From any point P of S drop the perpendiculars to the six edges of the strips, that is, the six sides of the triangles. By Viviani's theorem these perpendiculars sum by threes to the altitudes h and

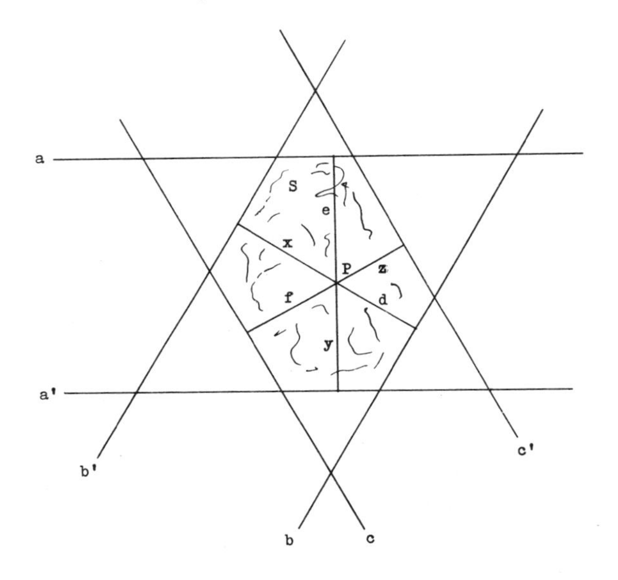

k of the two equilateral triangles:

$$d + e + f = h \quad \text{and} \quad x + y + z = k.$$

However, the edges of each strip are parallel. Thus the perpendiculars go together in pairs to form three straight segments of lengths $x + d$, $y + e$, and $z + f$. Since the width of a minimal strip is not greater than 1 unit, each of these segments is no longer than 1 unit. The sum of the lengths of all three segments, containing all six perpendiculars, then, is not greater than 3. This gives

$$h + k \leqq 3.$$

As a result, not both altitudes h and k can exceed $3/2$. Thus at least one of our equilateral triangles must have an altitude which is no greater than $3/2$. This implies a side no greater than

$$\frac{2}{\sqrt{3}} \cdot \frac{3}{2} = \sqrt{3}.$$

The conclusion follows.

Exercises

1. Angles ABC and CBD are each $60°$. From the point P anywhere in the angle ABC perpendiculars PX, PY, PZ are drawn to AB, BC, BD, respectively. Prove that $PZ = PX + PY$. (Hint: Viviani's theorem.)

2. Given a point P in the plane. Construct an equilateral triangle ABC such that P occurs inside it and is x units from A, y units from B, and z units from C. (Hint: Rotate completed figure through 60 degrees about a vertex.)

3. $ABCD$ is a parallelogram in which no angle is $60°$ or $120°$. Equilateral triangles ADE and DCF are drawn outwardly on sides AD and DC. Prove that ΔBEF is equilateral.

4. D is the midpoint of the side BC of equilateral triangle ABC. Outwardly on BD and DC equilateral triangles BDE and DCF are drawn. Prove that AE and AF trisect BC.

5. Outwardly on the sides of ΔABC are drawn any three triangles ABD, BCE, CAF such that $\angle D + \angle E + \angle F = 180$ degrees. Prove that the circumcircles of the latter three triangles are concurrent.

Prove also that the triangle formed by the centers of these circles has angles equal to $\angle D$, $\angle E$, $\angle F$. (Hint: The line of centers of two circles is perpendicular to their common chord.) Napoleon's Theorem now follows immediately.

6. Napoleon's Theorem is valid even if the three equilateral triangles are drawn inwardly instead of outwardly. Prove this.

7. In Napoleon's Theorem, the equilateral triangle of centers is called the "outer" Napoleon triangle if the equilateral triangles are drawn outwardly, and the "inner" Napoleon triangle if they are drawn inwardly. Prove that the difference of the areas of the outer and inner Napoleon triangles is just the area of the original triangle.

8. Prove that the outer and inner Napoleon triangles have the same center. (See exercise 7.)

9. A regular hexagon H of side $2/\sqrt{3}$ is drawn in the plane. A set of disks of unit radius is placed on the plane so that the center of each disk lies in H. Prove that the entire set of disks can be pinned

to the plane by placing three thumbtacks at the vertices of an appropriate equilateral triangle of unit side.

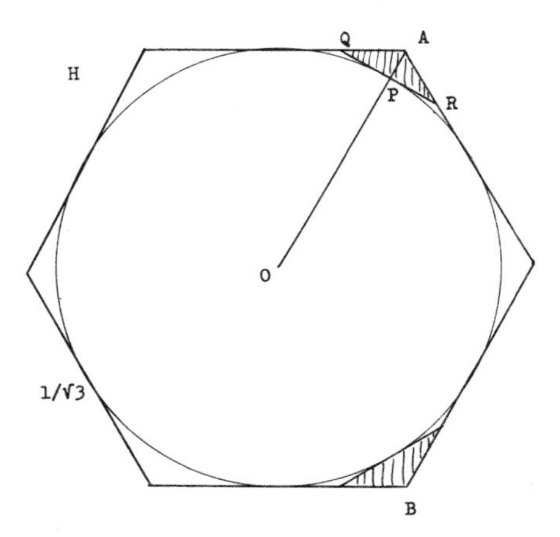

10. If a set S of points of diameter 1 can be covered by a regular hexagon H of side $1/\sqrt{3}$, show that the set S can also be covered by H with two alternate corners cut off as follows:

(i) Draw the incircle of H; let O denote its center; (ii) let OA (A the vertex of H at which the corner is to be cut off) meet the circle at P; (iii) let the tangent to the circle at P meet H at Q and R; (iv) cut off the corner AQR; (v) similarly at vertex B.
(This diminished hexagon is known as Pál's universal cover of plane sets of diameter 1, after the Hungarian mathematician J. Pál who proposed it in 1920.)

11. A point P is chosen at random inside an equilateral triangle. What is the probability that the perpendiculars to the sides of the triangle from P can be arranged to form a triangle?

12. On the sides of an arbitrary convex quadrilateral $ABCD$ equilateral triangles are drawn alternately outwardly and inwardly. Prove that the four new vertices obtained determine a parallelogram.

13. Triangle ABC is equilateral. A continuous line, not necessarily straight, is drawn from some internal point of AB to some

internal point of AC such that it bisects the area of ΔABC. Determine where this line should be drawn and what shape it should have in order for it to be as short as possible.

14. AB and CD are parallel lines which are cut by a transversal at E and F, respectively. The trisectors of angles FEB and EFD intersect to give the points P and Q (let the trisectors adjacent to EF meet at P, and the other two at Q). PG is drawn parallel to FQ to meet EQ at G, and PH is drawn parallel to EQ to meet FQ at H. GH extended in both directions meets AB at J and CD at K. Show that G and H trisect JK.

References and Further Reading

1. Coxeter and Greitzer, Geometry Revisited, Random House, New York, New Mathematical Library Series, vol. 19.

2. H. Meschkowski, Unsolved and Unsolvable Problems in Geometry, Oliver and Boyd, 1966.

3. H. S. M. Coxeter, Introduction to Geometry, Wiley, New York, 1961.

CHAPTER 4

THE ORCHARD PROBLEM

In this essay we consider an intriguing problem and the beautiful mathematics involved in its solution. We note that the points in a coordinate plane which have both coordinates integers are called lattice points.

THE ORCHARD PROBLEM: A tree is planted at each lattice point in a circular orchard which has center at the origin and radius 50. (All trees are taken to be exact vertical cylinders of the same

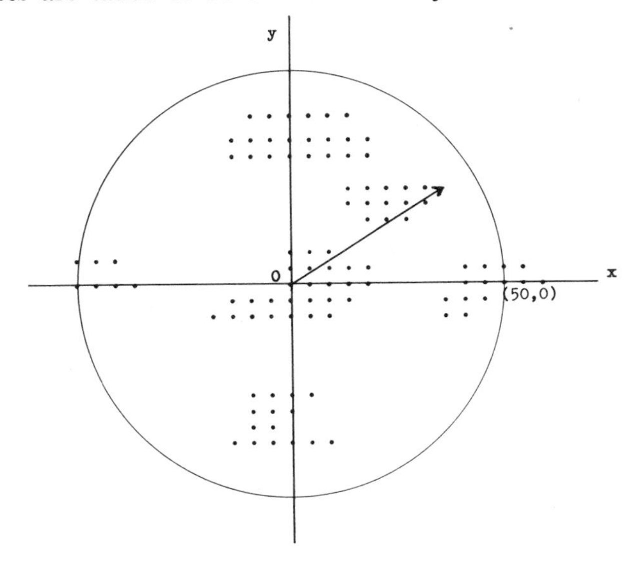

radius.) If the radius of the trees exceeds $\frac{1}{50}$ of a unit, show that from the origin one is unable to see out of the orchard no matter in what direction he looks; show, however, that if the trees are shrunk to a radius less than $1/\sqrt{2501}$, one can see out if he looks in the right direction.

The second part of the problem is quite easy. The solution of the first part turns on an interesting theorem of Hermann Minkowski (1864–1909), a close friend of Hilbert. In order to prove this theorem, we establish first a result called Blichfeldt's Lemma, after an American mathematician.

BLICHFELDT'S LEMMA. *Suppose we are given a plane region R which has an area in excess of n units, n a positive integer. Then, no matter where R occurs in the plane, it is always possible to translate it (i.e., to slide it without turning it) to a position where it covers at least $n + 1$ lattice points.*

For example, if R has an area of $8\frac{1}{4}$, then it can be translated to cover at least 9 lattice points.

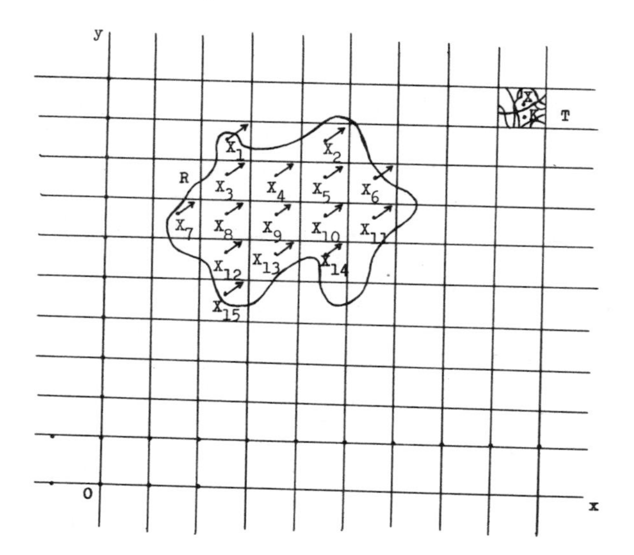

The lines whose equations are $x = a$ and $y = b$, a and b integers, we shall call "lattice lines" since their intersections are the lattice points. Let us cut the plane along each lattice line, thus chopping it up into squares of unit size. This also cuts our region R into pieces. Let us suppose that the region R is painted red and that the rest of the plane is unpainted. Some of the unit squares, then, may be all red, some partly red, and others completely unpainted.

Let us take all the squares which have any red at all on them and pile them up, without turning them, one on top of the other, in some distant square T. Consider now any point K of the base T. On top of T is piled layer after layer, each one covering the point K with some point of the layer. Sometimes K may occur under a red point in the layer, sometimes under an unpainted point. Among the vertical column of points that pile up on top of K we are interested in how many are red.

It is our claim that some point of the base T must get covered with a red point by at least $n + 1$ different layers. (Recall that the integer n enters our considerations through the fact that the area of R exceeds n units.) To prove this, consider the contrary. Suppose that no point of T lies under more than n red points in its column. Some points may occur at the bottom of a column containing exactly n red points, some fewer, but none have more than n red points.

Now let us calculate the amount of red paint in the pile. The area of the base T is one square unit, and even if every point of T had the maximum number, n, of red points in the column above it, this would provide only enough red paint to give the square n coats. (Each point of T could be given n coats with the n red points above it.) Thus, at most there is only enough paint in the pile to cover n units of area. But all of R is present in the pile, and it has an area greater than n units. Thus there must be enough paint in the pile to cover more than n units, contradiction. Consequently, some point X of T is covered at least $n + 1$ times with red points of R.

Now drive a needle straight down through all the layers above the point X. This marks a point in each layer, and, because of the above, it must occur in the red part of at least $n + 1$ layers. Let us denote these red points $X_1, X_2, \ldots, X_m$. Here m is at least $n + 1$.

Finally, return all the squares to their original places in the plane, thus reconstructing R.

Now each point X_i occurs in its square in the same relative position. Consequently, any translation of R which carries one X_i to cover a lattice point will also carry every other X_i to cover a lattice point. But each X_i is red, and is therefore a point of R. And there are at least $n + 1$ of them. Such a translation, then, carries R to cover at least $n + 1$ lattice points. (QED)

Blichfeldt's Lemma is very important in our proof of Minkowski's Theorem. However, it does not figure directly. Rather, we use an easy corollary which arises in the case $n = 1$.

COROLLARY. *If R is a plane region with area exceeding* 1, *then some pair of distinct points A and B of R have a run and a rise which are both integers. (The run between the points (x_1, y_1) and (x_2, y_2) is $x_2 - x_1$ and the rise is $y_2 - y_1$.)*

Notice that there is no claim that the points A and B, themselves, are lattice points. The corollary holds whether or not R covers any lattice points at all.

We know by Blichfeldt's Lemma that R can be translated so that (at least) $n + 1 = 2$ of its points, say A and B, are carried onto lattice points A_1 and B_1. Since the coordinates of lattice

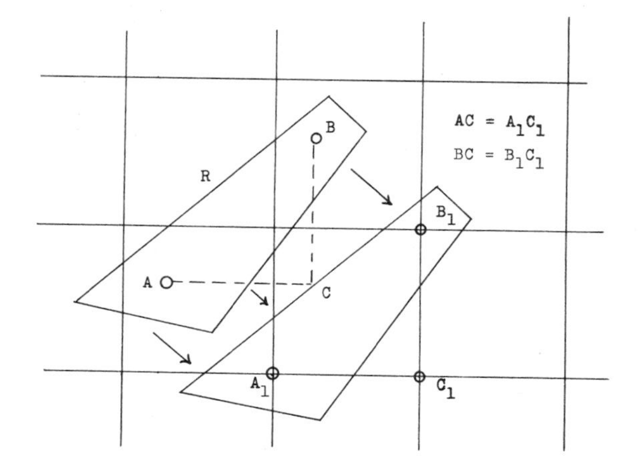

points are integers, the run and rise between any two of them are integers. However, a translation, changing only the position and not the direction of a line, does not change runs or rises. The run and rise between A and B before they are moved by the translation are exactly the same as the integral run and the integral rise between the lattice points A_1 and B_1 onto which they are carried.

(QED)

Now we are ready to prove an intuitively obvious, but not logically evident, theorem of Minkowski.

MINKOWSKI'S THEOREM. *A plane, convex region with area exceeding* 4, *which is symmetric about the origin, covers a lattice point besides the origin.*

(A convex set contains the entire segment AB for every pair of points A and B that it contains.)

If a region is symmetric about the origin, then for each point P that it contains it contains also the point P' which is obtained from P by reflection in the origin, that is, by sending it through the origin the same distance on the other side. If P has coordinates (x, y), then P' has coordinates $(-x, -y)$. We observe, then, that if the region covers one additional lattice point besides the origin, it must cover also a second lattice point which is symmetrical with the origin.

Let R denote the given region, and let us shrink R towards the

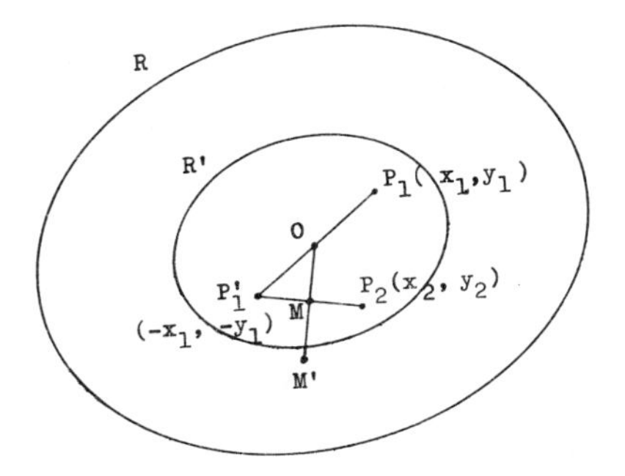

origin O until it is just half as large in every direction; that is, draw every point of R towards the origin until its new position, while remaining in line with the old, is just half as far from the origin. This transformation is called the dilatation with center O and ratio 1/2. Suppose R is taken into the region R' by this dilatation. Now dilatations carry lines into parallel lines (in the diagram, let A', B' denote the images of A, B; then $A'B'$ is parallel to AB, and the points C of the segment AB go into the points C' of $A'B'$). As a result, a dilatation does not change the size of the angles in a figure, and therefore it does not alter the shape of a figure. Consequently, R' has the same shape as R, only it is smaller. That is to say, R' also is a plane, convex region which is symmetric about the origin. It is symmetric about the origin O because O is the center of symmetry of R and also the center of the dilatation.

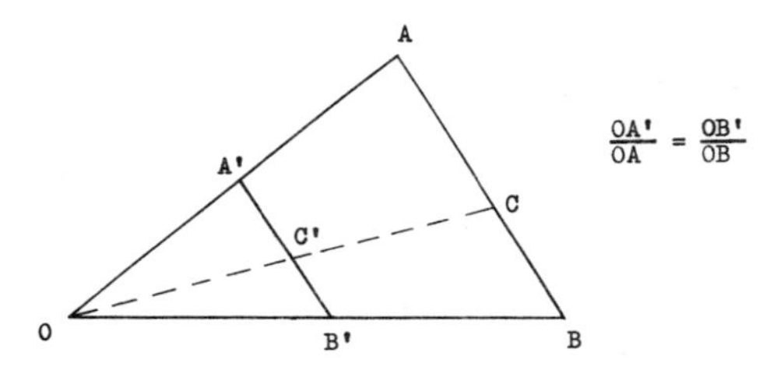

But what is the area of R'? Its width in every direction is just half what it used to be. If R were a rectangle, then R' would be a rectangle with half the length and half the width. In all cases, however, reducing the linear dimensions of a plane figure in the ratio $1:2$ reduces the area in the ratio $(1:2)^2 = 1:4$. Since R began with an area exceeding 4, we conclude that R' still has an area exceeding 1. Thus we may apply the Corollary to Blichfeldt's Lemma.

Accordingly, R' contains two points $P_1(x_1, y_1)$ and $P_2(x_2, y_2)$ whose run and rise, $x_2 - x_1$ and $y_2 - y_1$, are both integers. Because the origin O is the center of R', the point $P_1'(-x_1, -y_1)$ which is symmetric to P_1 in the origin must also lie in R'. Thus P_1' and P_2 are two points which belong to R'.

Now we use the fact that R' is convex. By definition, this assures us that every point of the segment $P_1'P_2$ belongs to R'.

In particular, its midpoint $M((x_2 - x_1)/2, (y_2 - y_1)/2)$ occurs in R'.

Now let us reverse our dilatation by subjecting R' to the dilatation with center O and ratio 2:1. This stretches R' back to its original state, namely R. Under this dilatation, every point of R' is moved to a point twice as far from the origin. The point M, then, is carried into the point $M'(x_2 - x_1, y_2 - y_1)$. Thus M' is a point of R. But, by Blichfeldt's Corollary, M' is a lattice point. And it is not the origin, itself, for that would mean that P_1 and P_2 coincide, in contradiction to Blichfeldt's Corollary, which asserts that they are different points. (QED)

Now let us attack the first part of the orchard problem. We wish to show that if the radius r of the trees exceeds $\frac{1}{50}$ of a unit, then there is no way to see out of the orchard from the origin. Let AOB denote an arbitrary diameter of the orchard. Suppose the radius of the trees is $\frac{1}{50} + q$. We observe that if r is greater than 1/2, the trees will be growing into each other. Thus, while q is positive, it is not a very large number. Now let p denote any number which is bigger than 1/50 but less than r, say $\frac{1}{50} + \frac{1}{2}q$. At A and B on the boundary of the orchard construct tangents, and cut off along these tangents in both directions points C, D, E, F at a distance p from A and B.

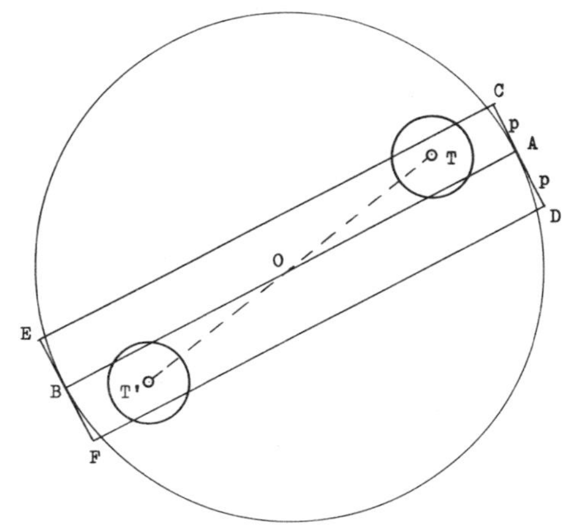

Thus a rectangle $EFDC$ is determined which has the origin O as center. (The diagram greatly exaggerates the size of this rectangle.) The length of the rectangle is $FD = AB = 100$, and the width is $2p$, giving an area of $200p$. Because p exceeds $1/50$, this area exceeds 4. And $EFDC$ is a plane, convex region which is symmetric about the origin. By Minkowski's theorem, then, our rectangle contains some lattice point T other than the origin. The tree which is planted at T has radius r, which exceeds $p = CA$. Thus the tree at T extends far enough to cross the line OA, blocking one's view in this direction. By symmetry, the rectangle contains also the symmetric lattice point T', at which is planted a tree that blocks the view along OB. Thus it would appear that we may conclude that one cannot see out from the origin. However, there is a difficulty in our argument which must be overcome (do you see what it is?).

A very small part of the rectangle sticks out of the orchard at each corner. If the lattice point T happens to occur in one of these parts of the rectangle, then there is no tree planted there to block the view. We need to show that T does not occur in a part of $EFDC$ which lies outside the orchard. We proceed indirectly. For any point in the rectangle the maximum distance from the origin is the half-diagonal $OC = \sqrt{50^2 + p^2}$. Since $p < 1$, we have that $OT \leqq OC < \sqrt{2501}$. However, for T outside the orchard we have $OT > 50$. Hence

$$2500 < OT^2 < 2501.$$

If T is the lattice point (x, y), then $OT^2 = x^2 + y^2$, where x and y are integers, implying that OT^2 is an integer. But there is no integer between 2500 and 2501. Thus T cannot occur outside the orchard.

We complete the solution by showing that if the radius is diminished to anything less than $1/\sqrt{2501}$, one can see out of the orchard along the line joining the origin to the point $N(50, 1)$. The length of the segment ON is $\sqrt{2501}$. Since the lattice points occur in rows and columns, it is easy to see that the lattice point in the orchard which is nearest to the line ON is the point $M(1, 0)$, or the equally close $K(49, 1)$. Let L denote the point $(50, 0)$.

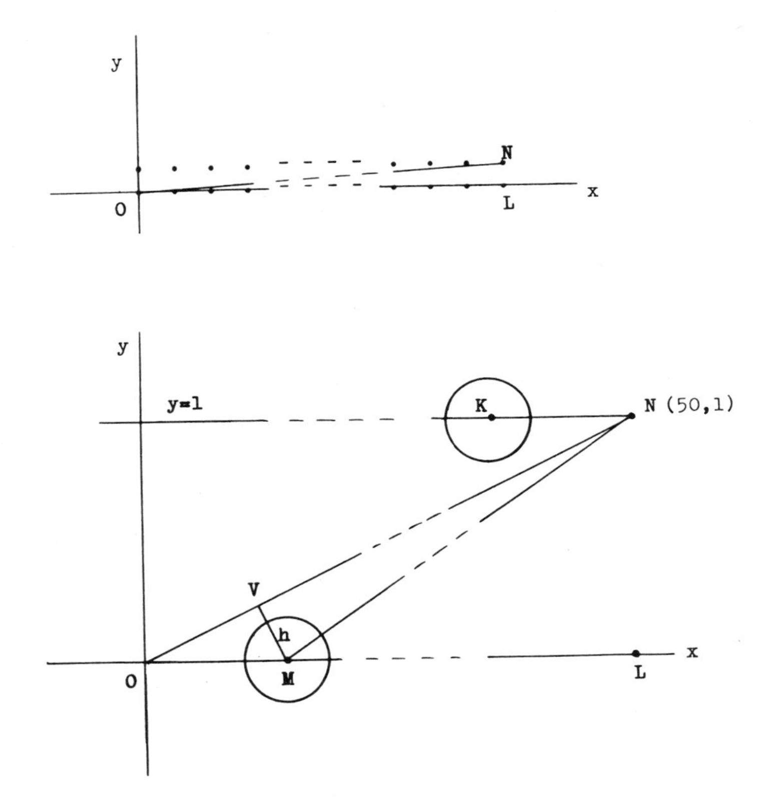

Now the area of ΔOMN can be obtained in two ways. First, it is given by $\frac{1}{2} \cdot OM \cdot LN$, which yields $\frac{1}{2} \cdot 1 \cdot 1 = \frac{1}{2}$. Using ON as base and the altitude $h = MV$, we also get $\frac{1}{2} \cdot ON \cdot h$, which yields $\frac{1}{2}h\sqrt{2501}$. Consequently, we have

$$\frac{1}{2}h\sqrt{2501} = \frac{1}{2}, \qquad \text{giving} \quad h = \frac{1}{\sqrt{2501}}.$$

Thus h exceeds the radius of the trees. As a result, the tree at M is not big enough to intersct the line ON; similarly, neither is the tree at K. Since the closest trees do not cross ON, then no tree blocks the view in this direction.

Exercises

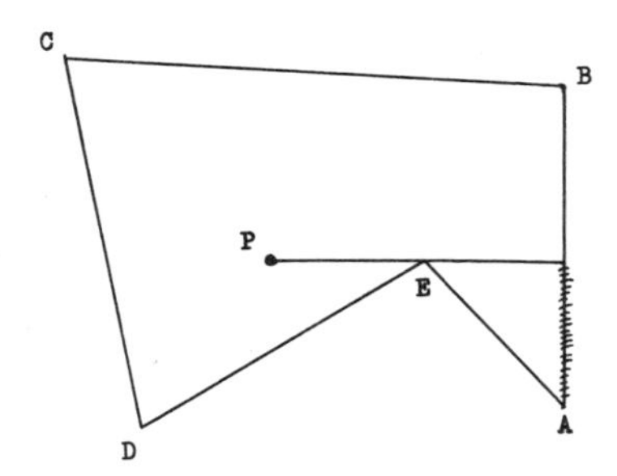

1. The figure shows a point P inside a polygon from which part of the side AB and all of AE are not visible. Construct a polygon G with a point P inside it such that no side of G is completely visible from P. Construct a polygon H with a point P outside it such that no side of H is completely visible from P.

2. Given a circle, center O, and a point P outside it, construct a straight line PQR (whenever possible), Q and R on the circle, with Q the midpoint of PR. (Hint: Consider the dilatation with center P and ratio 1/2.)

3. In a square lattice, show that (a) no matter what three lattice points are joined, an equilateral triangle never results, (b) no matter what five lattice points are joined, a regular pentagon never results.

4. Show that a convex region of area 1 can be covered by some parallelogram of area not greater than 2.

5. A circle C_0 of radius $R_0 = 1$ km. is tangent to a line L at Z. A circle C_1 of radius $R_1 = 1$ mm. is drawn tangent to C_0 and L, on the right-hand side of C_0. A family of circles C_i is constructed outwardly to the right side so that each C_i is tangent to C_0, L, and to the previous circle C_{i-1}. Eventually the members become so big that it is impossible to enlarge the family further. How many circles can be drawn before this happens? (See page 53.)

6. Look up pages 43 and 44 in Hilbert's *Geometry and the Imagination* for an application of Minkowski's theorem to the problem of approximating real numbers by sequences of rational numbers.

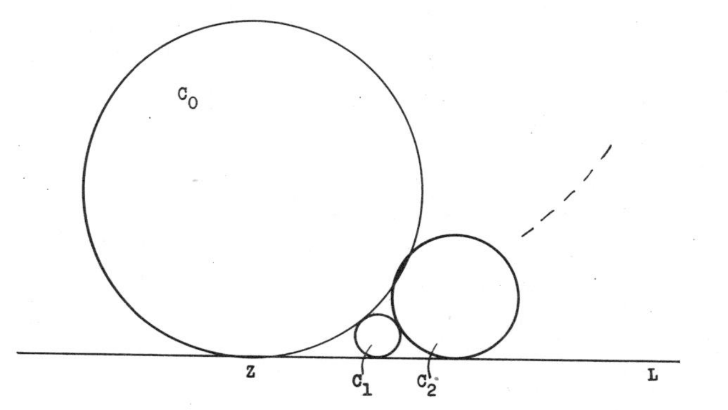

7. L. G. Schnirelman (Russian: 1905–1935) proved that some four points on a closed convex curve are the vertices of a square. Use this result to show that every convex curve of perimeter less than four can be positioned on a coordinate plane so as to cover no lattice points at all.

Reference and Further Reading

1. Yaglom and Yaglom, Challenging Mathematical Problems with Elementary Solutions, vol II, Holden-Day, San Francisco, 1967.

CHAPTER **5**

Δ-CURVES

1. Introduction. If we wished to know the width w of a plane figure Q in the direction of a line m, we would measure the width of the narrowest strip containing Q which runs in a direction perpendicular to m. Each edge of such a minimal strip is called a supporting line, and the two edges form a pair of parallel supporting lines of Q. If Q has no boundary (like the set of interior points of a circle) then supporting lines will not actually make contact with Q. In what follows, we consider all figures to possess their boundaries, in which case a supporting line does pass through at least one point of Q, and, discounting points of contact on the line, it has all of Q on the same side of it.

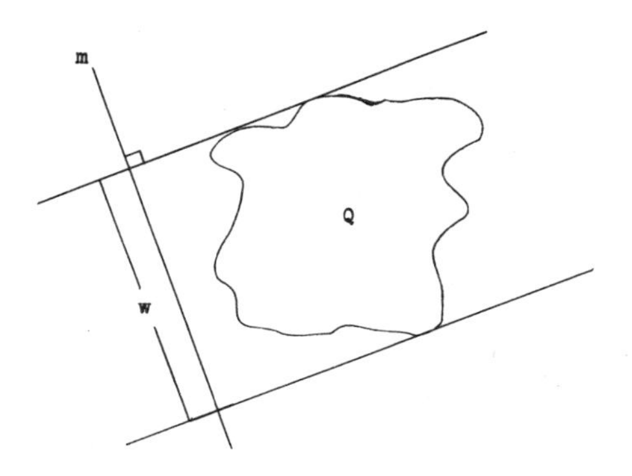

If the width of Q is the same in all directions, Q is said to be a curve of constant breadth, or constant width. Obviously, a circle is such a curve. However, other examples do not readily spring to mind. A so-called Reuleaux triangle is one example of such a figure. It is constructed by drawing arcs with centers at the vertices A, B, C of an equilateral triangle and radius equal to a side of the triangle. One of a pair of parallel supporting lines always passes through a corner and the other is tangent to the opposite arc, making the width constantly equal to the radius.

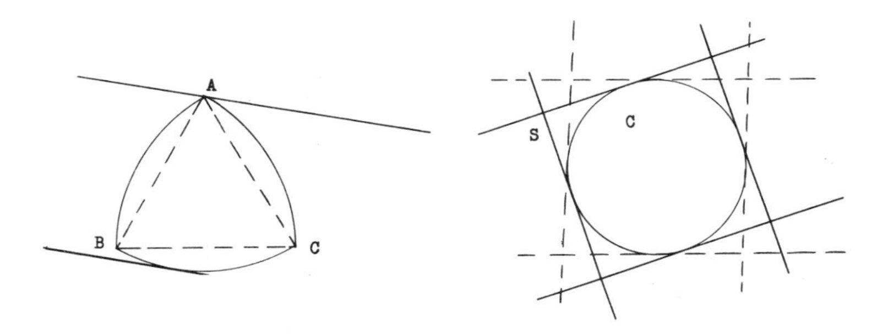

There are many curves of constant breadth. In fact, among these curves there is an infinity of different shapes. If one constructs perpendicular pairs of parallel supporting lines to a curve of constant breadth C, then C is enclosed in a square which has the same size no matter what the direction of the supporting lines. Thus, keeping C fixed, one could rotate around it a square S which continuously remains in contact with C along its four sides. Since motion is relative, we could instead rotate C inside S. Thus we may think of curves of constant breadth C as those curves which can be turned inside some square S while continuously making contact with each side. Curves of constant breadth have many interesting properties. One of the most exciting is Barbier's Theorem:

All the various curves with constant breadth h have the same perimeter πh.

An excellent account of these curves is contained in the book *Convex Figures* by Yaglom and Boltyanski (Holt, Rinehart & Winston).

However, our interest in this essay is in a different class of curves, the so-called Δ-curves (delta-curves). In contrast to curves of constant breadth, these are the curves which can be turned continuously inside an equilateral triangle (rather than a square).

2. Δ-Curves. Again the circle is the only Δ-curve that comes to mind at the outset. Again, however, there is an infinite variety of shapes among the Δ-curves. Besides the circle, the simplest is a lens-shaped curve called a Δ-biangle.

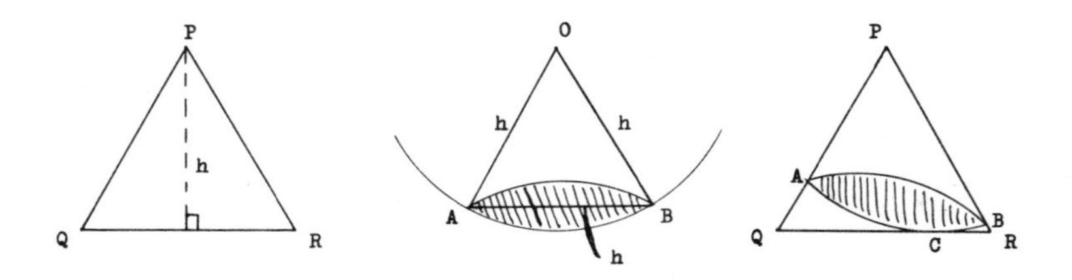

To construct a Δ-biangle for an equilateral triangle of altitude h, draw a circle with radius h and cut off a chord AB of length h (this produces an equilateral triangle with the radii to A and B). The Δ-biangle is obtained by reflecting the minor arc AB in the chord AB. In general, the Δ-biangle touches the sides of the equilateral triangle at the vertices A, B and at a point C on an arc. (However, C will coincide with A (or B) when A is at a vertex.)

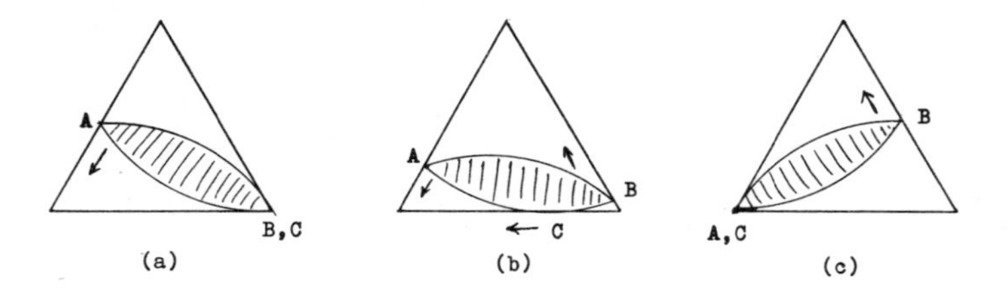

Diagrams (a), (b), and (c) attempt to illustrate how the Δ-biangle rolls around inside the triangle. Starting as in (a), imagine the Δ-biangle to roll on the base (as a tipped-up saucer would roll down on a table). During the motion, every point C of the arc AB becomes a point of contact with the base. And as the rolling takes place, the straight-line paths of the vertices A and B simply trace out parts of the other sides of the triangle.

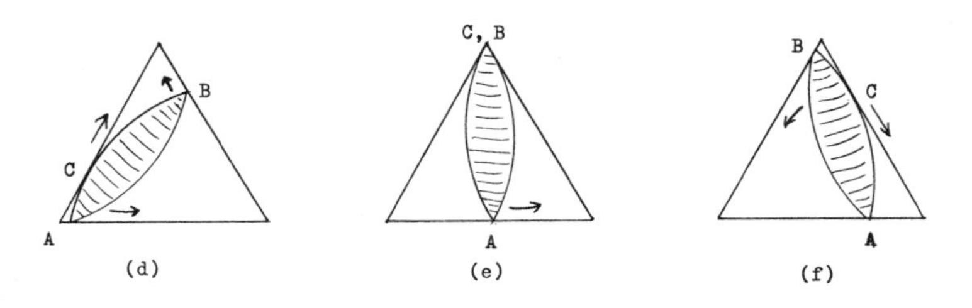

(d) (e) (f)

Diagrams (d), (e), and (f) show C travelling along the other sides of the triangle, giving a more complete picture of how the Δ-biangle rotates all the way around. (Somewhat surprisingly, we notice that C goes around all three sides while the Δ-biangle turns only half way around in the triangle.) Since it is far from obvious that a Δ-biangle is a Δ-curve, let us prove it.

We observe that an equilateral triangle can be circumscribed about any finite figure in the plane simply by bringing up to the figure three lines in the appropriate directions (at 120 degrees to each other). Once the direction of one side is decided upon, the directions of the other two sides are determined. However, one can choose to place the initial side in any direction one likes, implying that every finite plane figure possesses a continuous family of circumscribing equilateral triangles, ranging over all directions. What we need to show for our Δ-biangle is that all its circumscribing equilateral triangles are the same size. This implies that the Δ-biangle can be turned continuously inside any one of them (motion is relative). We show that they are all the same size by showing that the altitude is always h, the length of the chord AB of the Δ-biangle. Let PQR denote any circumscribing equilateral triangle, and let PS denote the altitude from P. We prove $PS = h$.

Let QR meet the arc of the Δ-biangle at C, and let O denote the center of the circle determining the arc ACB. Because QCR is a tangent, OC is perpendicular to QR and of length h equal to the radius. Also we have ΔAOB equilateral, making $\angle AOB = 60° = \angle APB$. Accordingly, $APOB$ is a cyclic quadrilateral in which the opposite angles at P and B are supplementary. However, the

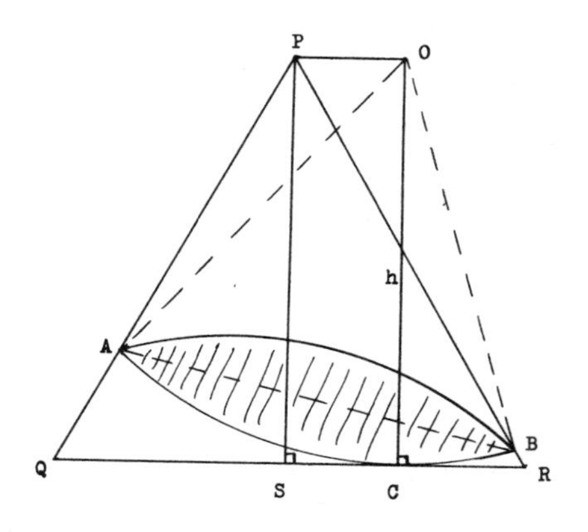

angle at B, $\angle ABO$, is $60°$ (in equilateral triangle ABO), implying that $\angle APO = 120°$. Consequently

$$\angle Q + \angle APO = 60° + 120° = 180 \text{ degrees},$$

making PO parallel to QR. Thus $POCS$ is a rectangle, giving altitude $PS = OC = h$ as desired.

The equilateral triangles circumscribed about a figure generally vary from direction to direction. The altitude h of the equilateral triangle drawn around a figure F so that an altitude occurs in a given direction AB is called the "height" of F in the direction AB. Accordingly, the Δ-curves are the curves of constant height.

Another example of a Δ-curve is the curve determined by the region common to the four quarter-circles drawn in a square with the vertices as centers and the side as radius.

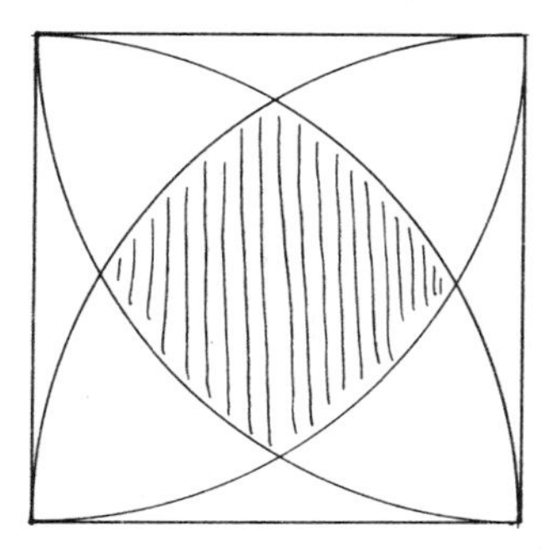

Δ-curves have many fascinating properties. We mention only two. At each position of a Δ-curve turning in an equilateral triangle, the perpendiculars to the sides at the points of contact are concurrent (at the "instantaneous center of rotation"). Secondly, all the various Δ-curves of height h have the same perimeter $\frac{2}{3}\pi h$ (Barbier's theorem for Δ-curves).

3. Related Curves. Every plane figure F has a continuous family of circumscribing rectangles, ranging over all directions. If all these rectangles are squares of the same size, F is a curve of constant breadth. Another interesting class of curves is obtained by relaxing this definition to the single requirement that all the rectangles be equal in size, whether or not they are squares. The curves of constant breadth are automatically included in this class. But it includes other curves, too. We interpret "equal in size" to refer to perimeter rather than area.

The simplest curve of this class, which is not also a curve of constant breadth, is the curve determined by the lens-shaped region common to two quarter-circles drawn in a square with a pair of opposite vertices as centers and with the side as radius. Let us prove that it does belong to our new class of curves.

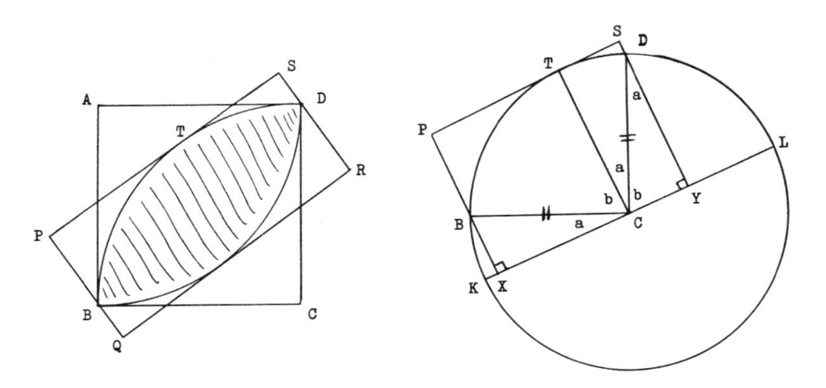

The following simple proof is due to J. Levitt of the Ontario Department of Education. Let $ABCD$ denote the defining square and let $PQRS$ denote an arbitrary circumscribing rectangle. Now the vertices B and D of the square provide two points of contact with the rectangle, and they divide its perimeter in half. We establish our result by showing that the half-perimeter $BP + PS + SD$ is constantly equal to the sum of two sides of square $ABCD$.

Let PS touch the curve in question at T, and let arc BTD be completed into a circle center C. Let KCL denote the diameter which is perpendicular to TC, and let PB and SD extended meet this diameter at X and Y. It follows easily, then, that triangles BXC and CDY are congruent, making $BX = CY$ and $DY = XC$. With $PX = TC = SY$, we have

$$PB + SD = (PX - BX) + (SY - DY) = (PX + SY) - (BX + DY)$$

$$= 2 \cdot TC - (CY + XC) = 2 \cdot TC - XY = 2 \cdot TC - PS.$$

Adding PS, we see that the half-perimeter is just $2 \cdot TC$, or twice the radius of the arcs, as desired.

A second curve of this class, which is also not a curve of constant breadth, is obtained from an equilateral triangle ABC as follows. Extend the altitudes outside the triangle to give points X, Y, Z such that $AX = BY = CZ =$ a side of the triangle. With centers X, Y, Z draw arcs to join the two opposite vertices. The resulting Reuleaux-triangle-type curve has all circumscribing rectangles with the same perimeter.

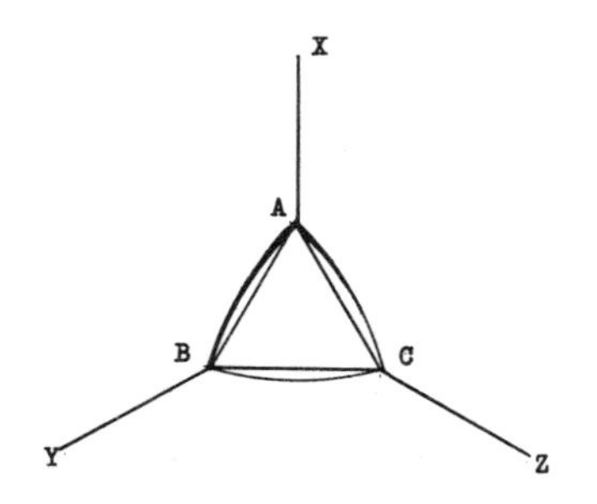

We leave these curves with the remark that of the various curves whose circumscribing rectangles have constant perimeter $4h$, all have the same perimeter πh (Barbier's theorem extends over this entire class of curves).

4. A Converse Problem. Finally, let us switch things around and consider the curves in which an equilateral triangle can be turned while maintaining continuous contact with the curve. Besides the obvious member of this class, the circle, only one other member is known. It is the curve determined by the region common to two equal circles, each of which passes through the center of the other. Let X and Y denote the centers and A and B the points of intersection. Then triangles AXY, BXY are clearly equilateral, and so is any triangle PQX where P and Q occur on the arc AYB and PQ has length equal to the radius. Thus ΔAXY can be swung about the point X to the position of ΔXYB, after which it can be pivoted back about the center Y (the new position of the vertex A) to its original position, having rotated through one-third of a revolution. Repeating this procedure, one can turn the triangle as far as desired.

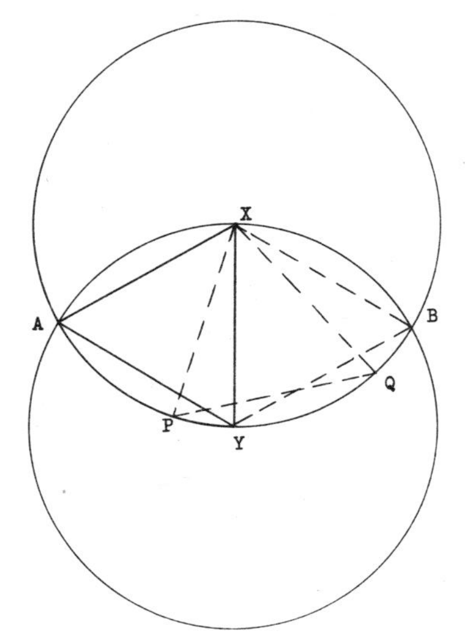

Exercises

1. Prove that the three perpendiculars to the sides of an equilateral triangle ABC at the points of contact U, V, W of a Δ-biangle in it are concurrent. (Hint: Suppose U, V, W occur on sides BC, AC, AB, respectively. Let the perpendiculars at V and W meet at P. Then the circle on AP as diameter passes through V and W. Show also that it passes through the center O of the edge WUV of the Δ-biangle. OVW is an equilateral triangle. Show that UP is perpendicular to BC.)

2. Let the points of contact of an equilateral triangle ABC of height h with a Δ-curve in it be U, V, W. On VW construct towards A an equilateral triangle VWX. Show that $XU = h$. (Hint: Draw a circle through XWV and make use of exercise 1. Assume the property of 1 generalizes to any Δ-curve, not just a Δ-biangle.)

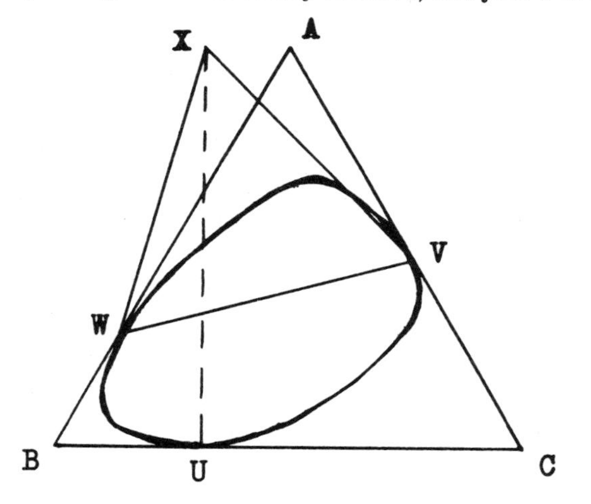

3. Let the sides of equilateral triangle ABC be folded out on the altitudes extended to give points X, Y, Z, as shown. With centers X, Y, Z, draw arcs to join B and C, A and C, A and B, respectively. Prove that the curvilinear triangle ABC thus obtained has the property that all its circumscribing rectangles have the same perimeter. (See figure on page 63.)

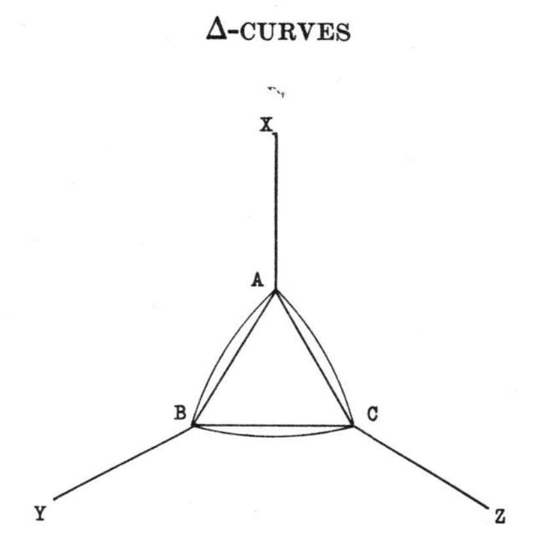

4. Show that every circumscribing rectangle of a square is a square.

5. Semi-circles are constructed outwardly on the sides of a right-angled triangle. A circumscribing rectangle is constructed around the figure so that its sides are parallel to the sides about the right angle of the triangle. Prove that the rectangle is a square.

6. Prove that it is always possible to circumscribe about a given convex figure F of area 1 a parallelogram of area less than or at most 2.

Reference and Further Reading

1. Yaglom and Boltyanski, Convex Figures, Holt, Rinehart & Winston, New York, 1961.

CHAPTER **6**

IT'S COMBINATORICS THAT COUNTS!

In this essay we consider a variety of short topics on the border between arithmetic and geometry. After a few problems involving the parity of numbers (whether odd or even), we turn to the remarkable story of Clifford W. Adams.

1. Parity.

(a) THE NUMBER OF DIVISORS OF A POSITIVE INTEGER. The brief table given lists for a few values of the positive integer n the number of divisors of n. We denote this number by $d(n)$. For example, the value of $d(4)$ is 3 because there are three divisors of 4, namely 1, 2, and 4.

n	1	2	3	4	5	6	7	8	9	10	11	12	13	14	15	16	17	18	19	20	$\cdots$
$d(n)$	1	2	2	3	2	4	2	4	3	4	2	6	2	4	4	5	2	6	2	6	$\cdots$

Looking over the values of $d(n)$, we are struck by the frequency with which $d(n)$ is an even number. As far as our table goes, the values of n for which $d(n)$ is odd are only $n = 1, 4, 9, 16$. Thus we have the first hint of a surprising theorem.

A SURPRISING THEOREM. *The number of divisors $d(n)$ of the positive integer n is an odd number if and only if n is a perfect square.*

The proof follows the simple observation that the divisors of n generally go together in pairs. We note that if $ab = n$, both a and b are divisors of n. Since $\sqrt{n}\cdot\sqrt{n} = n$, if $a < \sqrt{n}$ it must be that

$b > \sqrt{n}$. Thus the divisors of n which are smaller than $\sqrt{n}$ pair off with those which exceed $\sqrt{n}$. Generally these pairs constitute all the divisors of n, and $d(n)$ is even. If, and only if, n is a square do we have $\sqrt{n}$ an integer, yielding an extra divisor and making $d(n)$ odd.

(b) GOMORY'S THEOREM. Next we consider a challenging problem concerning a chessboard. One need not know anything about chess in order to enjoy it. We introduce it with an old chestnut.

Suppose we are given an ordinary 8×8 chessboard and 32

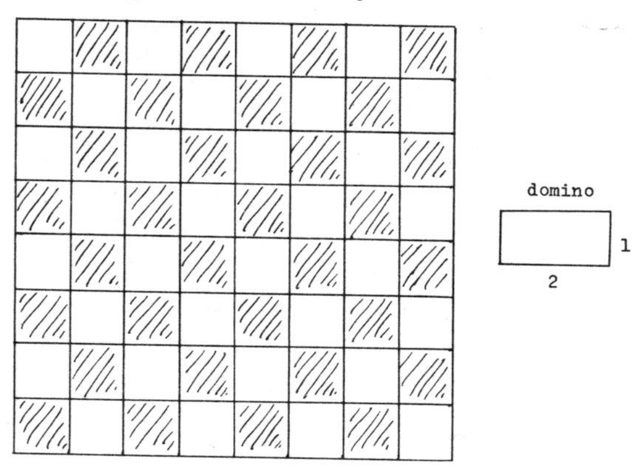

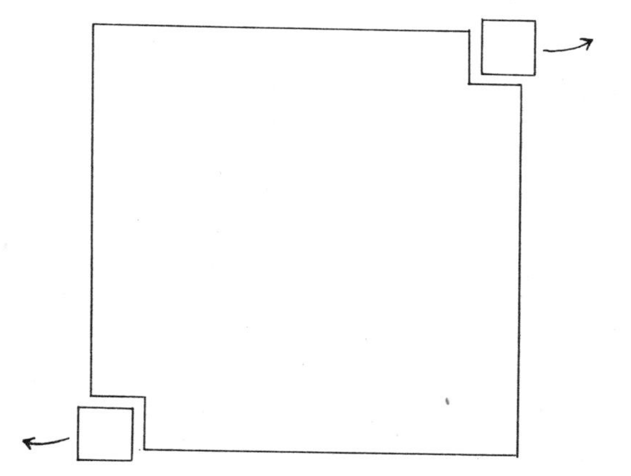

dominoes of dimensions 2×1. Obviously the dominoes can be arranged on the board to cover it completely. Now, two opposite corner squares are cut away and one domino is discarded. The problem is to determine whether or not the remaining 31 dominoes can be arranged on the reduced board so as to cover it exactly.

The answer is that it is impossible to cover the reduced board with the 31 dominoes. The full board has 32 black and 32 white squares. Observing that opposite squares are the same color, say black, the reduced board has 32 white but only 30 black squares. Now each domino covers a pair of adjacent squares, one of which must be black and the other white. Consequently, 31 dominoes can cover only 31 white squares, not the required 32 white squares. Thus the covering is hopeless.

It was pretty sneaky of the poser of the problem to reduce the board by referring to a pair of opposite corners when all he really wanted to do was to get rid of two squares of the same color. Of course, deleting two squares of the same color from any places on the board leaves the same impossible problem. The parity is thrown off. Our main question asks whether or not a covering is possible when the board is reduced by removing one white and one black square. It is not difficult to see that at least sometimes a covering can be affected (for example, when adjacent corners are deleted). However, it is with considerable appreciation that we acknowledge the following theorem due to Ralph Gomory, a mathematician with International Business Machines.

GOMORY'S THEOREM. *Regardless of where one white and one black square are deleted from an ordinary chessboard, the reduced board can always be covered exactly with 31 dominoes of dimensions 2×1.*

Proof. Place a three-tongued fork and a four-tongued fork on the board as shown in the diagram. The maze-like effect of this is to put the squares of the board in a cyclic order. One can walk around the maze through all the squares and wind up where he began.

Suppose now that some black square A and some white square B are deleted. Observe that in the cyclic order the squares alter-

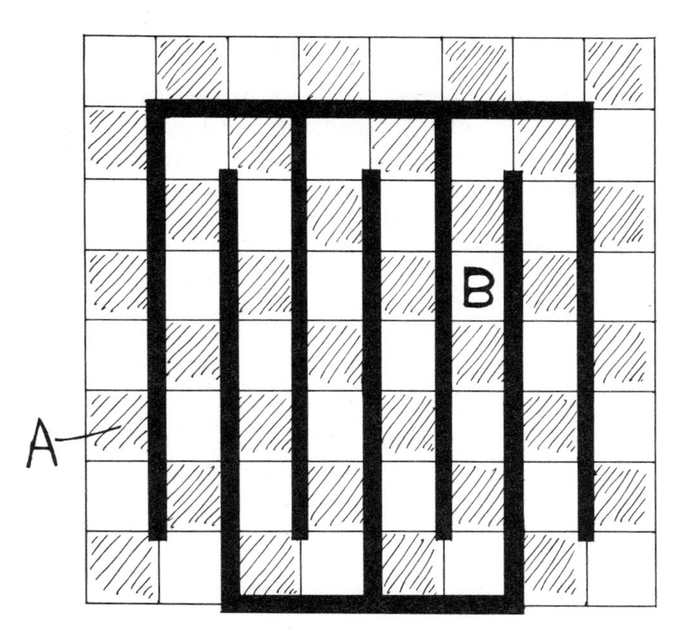

nate in color. As a result, the number of squares between a black square and a white square along the path of the maze is always an even number. Accordingly there is exactly enough room for a whole number of dominoes between A and B, if only we are able to fit them in around the twisted path in the maze. The only troublespots are at the corners. But, because the dominoes can be turned to run across or up-and-down the board, we can always get around a corner without leaving any gaps. Following the two paths from A to B along the maze, we are therefore able to cover the reduced board as desired.

(c) THE RHOMBIC DODECAHEDRON. As discussed briefly at the beginning of Dirac's theorem in the essay on Louis Pósa, in 1857 the Irishman Sir William Rowan Hamilton invented the game of travelling around the edges of a graph from vertex to vertex. We note that the object of the game was to find a path along the edges which passed through every vertex exactly once. Actually the edges and vertices of the game he proposed were those of a regular dodecahedron, and there are many solutions to the game. In this section we show by a simple parity argument that there is no Hamiltonian path on the surface of a rhombic dodecahedron. The proof is due to H. S. M. Coxeter, a distinguished Canadian geometer.

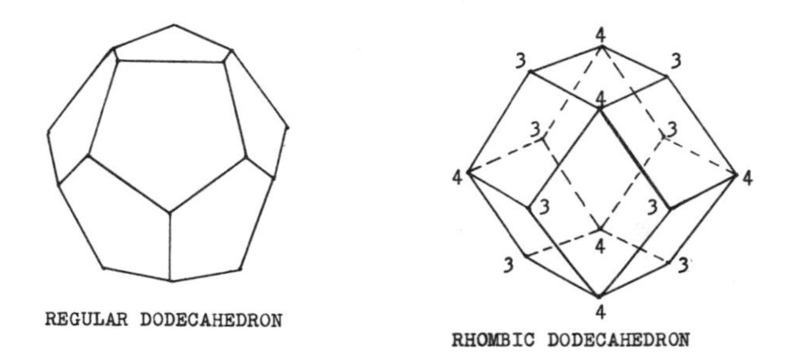

REGULAR DODECAHEDRON

RHOMBIC DODECAHEDRON

Observe that the valence of each vertex is either 3 or 4 (the valence of a vertex is simply the number of edges which run into it). Furthermore, each vertex of valence 3 is surrounded by vertices of valence 4, that is, each of its three edges runs to a vertex of valence 4. Conversely, each vertex of valence 4 is surrounded by vertices of valence 3. Consequently, a Hamiltonian path would have to alternate between vertices of valence 3 and vertices of valence 4. In order to carry through all 14 vertices, such a path would have to contain 7 vertices of each valence. However there are only 6 vertices of valence 4. Thus no Hamiltonian path is possible.

(d) THE JAILER PROBLEM. According to the terms of a partial amnesty, a jailer made n passes along a row of n cells, originally locked, as follows:

In the first pass, he turned every lock (i.e., he opened them all); in the second pass, he turned every other lock, beginning with cell 2; in the third pass, he turned every third lock, beginning with cell 3; . . . .
. . . .

in the kth pass, he turned every kth lock, beginning with cell k;

. . . .

in the nth pass, he turned every nth lock, beginning with cell n.
After all n passes, which cells remain unlocked?

From an investigation of the fate of the first dozen cells one can easily guess the answer and then try to prove it. However, we are happily in a position to dispose of the matter through the following simple approach. Consider the mth cell. It is locked to begin with. It has its lock turned on the first pass, and if m is divisible by 2, also on the second pass. Again, if m is divisible by 3, its lock is turned on the 3rd pass, and so on. That is, its lock is turned once for each divisor of m. In order to be one of the open locks at the end, it must be turned an odd number of times. Thus cell m remains open if and only if m has an odd number of divisors, that is, if and only if m is a perfect square.

2. Clifford W. Adams. The recreational side of mathematics has long been of popular interest to both professional and amateur mathematicians. The subject of magic squares has an extensive history containing many striking results and clever proofs. As everyone knows, in a magic square the numbers along each row, column, and diagonal add up to the same total. *In 1910* Clifford W. Adams got interested in seeing if he could find a "magic hexagon", that is, an arrangement of the numbers 1, 2, 3, 4, 5, 6, 7, with six of them around one in the center, which provides the same sum along each line of numbers. From the diagram we see immediately that no such arrangement is possible: $x + y$ would have to be the same as $x + z$, implying $y = z$.

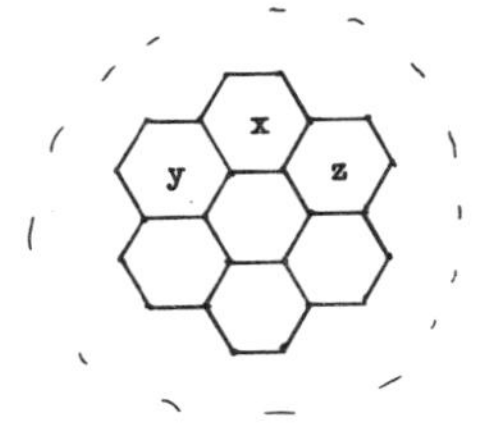

However, each number is to be used exactly once. So he decided to try for a two-layer hexagon by putting another twelve cells around the outside and using the numbers 1, 2, 3, $\cdots$, 19.

Clifford W. was a clerk for the Reading Railroad Company, and he worked at the hexagon in his spare time. He tried and he tried, but to no avail. He even had 19 little hexagonal tiles numbered 1 through 19 so he could carry out his investigations more readily. Some time later he retired and thus he was able to spend much more time searching for the hexagon. He continued to try, but he got nowhere. It was getting frustrating and exhausting. Finally, he got sick and had to go into the hospital. Naturally he took his numbered tiles with him.

One day when he was recuperating from his operation he was playing with his tiles and, lo and behold, he actually discovered a magic hexagon, . . . in the year 1957! Not wishing to forget this wonderful array, he copied it down on a piece of paper. But when he was released from the hospital, somewhow, in the shuffle, he lost the paper.

And so he tried to reconstruct the thing on his own. He tried and he tried, but to no avail. But he didn't give up. He continued to try for another five years until in December, 1962, what do you think happened? He found the paper! Here is his magic hexagon; each line adds up to 38.

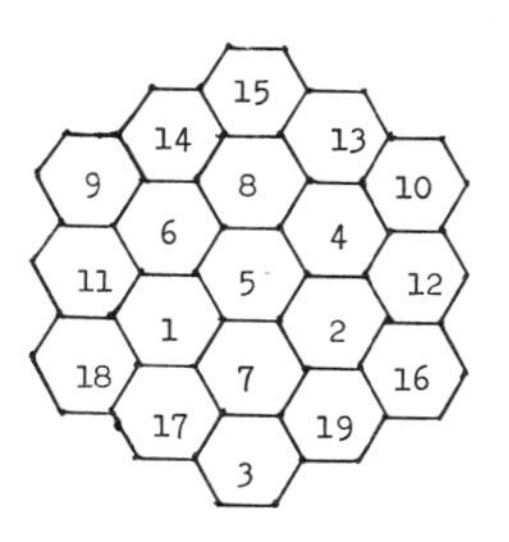

Now such a discovery should not go unnoticed. Adams sent his hexagon to Martin Gardner, a well-known columnist of Scientific American. Gardner knew that there are 880 different ways to arrange the numbers 1, 2, 3, . . ., 16 into a 4×4 magic square, and he reckoned that Adams' hexagon would be one among many. However, he could find no mention of such a thing anywhere in mathematical literature. He then communicated the matter to Charles W. Trigg, an ardent problemist and recreational mathematician from Los Angeles. Trigg went to work on it, and he was able to prove that of all the infinity of hexagonal arrays, involving any number of layers and using the numbers 1, 2, 3, . . ., as far as necessary, the one and only one which is a magic hexagon is Clifford W. Adams' hexagon! A splendid proof of this is due to Frank Allaire, who produced it in 1969 while a second-year student at the University of Waterloo. I am not familiar with Trigg's argument but I am led to believe that, except for Allaire's use of a computer, it is the same clever piece of reasoning. Allaire proceeds as follows.

Let n denote the number of layers, or rings, of cells around the central core. For $n = 0$, we have just the central cell, which is trivially magical. For $n > 0$, we show that if an array is magical, then $n = 2$. We begin by noting that every ring is in the shape of a regular hexagon, with a cell at each vertex, and with each edge of the nth ring ($n > 0$) containing $n - 1$ cells between its vertices. Thus each ring contains six more cells than the previous ring. Ac-

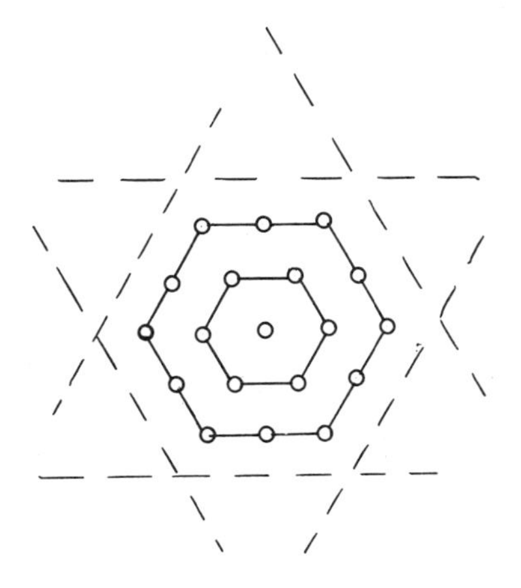

cordingly, the number of cells altogether in an n-ring array is the sum

$$1 + (6 + 12 + \cdots \text{ to } n \text{ terms})$$
$$= 1 + (6 + 12 + \cdots + 6n) = 1 + 6(1 + 2 + \cdots + n)$$
$$= 1 + 6 \cdot \frac{n(n+1)}{2} = 3n^2 + 3n + 1.$$

Next we observe that the hexagonal rings are concentric and have their edges parallel. The "lines" of cells which yield the constant sum lie in these three directions. In each of these directions there are $2n + 1$ such "lines" (there is the line through the middle and n lines on each side of it from the n rings).

Suppose the constant sum along a line is s. The $2n + 1$ s's one gets by adding along the $2n + 1$ lines in a given direction collectively counts the numbers in all the cells of the array. Thus the sum of the numbers in all $3n^2 + 3n + 1$ cells is $(2n + 1)s$. But these numbers are

$$1, 2, 3, \ldots, 3n^2 + 3n + 1, \text{ with sum } \frac{(3n^2 + 3n + 1)(3n^2 + 3n + 2)}{2}.$$

Consequently,

$$(2n + 1)s = \frac{(3n^2 + 3n + 1)(3n^2 + 3n + 2)}{2},$$

or

$$2(2n + 1)s = (3n^2 + 3n + 1)(3n^2 + 3n + 2).$$

Accordingly,

$$2n + 1 \mid (3n^2 + 3n + 1)(3n^2 + 3n + 2).$$

Now

$$3n^2 + 3n + 1 = (2n + 1)(n + 1) + n^2,$$
$$\text{and} \quad 3n^2 + 3n + 2 = (2n + 1)(n + 1) + n^2 + 1.$$

Multiplying gives a multiple of $2n + 1$ plus the term $n^2(n^2 + 1)$.

Thus

$$2n + 1 \mid n^2(n^2 + 1).$$

Surprisingly, however, the numbers $2n + 1$ and n^2 are always relatively prime. To prove this, let d denote their greatest common divisor. Then

$$d \mid 2n + 1 \quad \text{and} \quad d \mid n^2.$$

These easily give

$$d \mid (2n + 1)^2 = 4n^2 + 4n + 1,$$

$$d \mid 2(2n + 1) = 4n + 2,$$

and

$$d \mid 4n^2.$$

The latter two facts yield $d \mid 4n^2 + (4n + 2)$.
Combining with the first fact, we get

$$d \mid (4n^2 + 4n + 2) - (4n^2 + 4n + 1) = 1,$$

implying that $d = 1$.
From $2n + 1 \mid n^2(n^2 + 1)$, then, it follows that $2n + 1 \mid n^2 + 1$.

But $\quad n^2 + 1 = \frac{1}{4}[(2n + 1)(2n - 1) + 5]$,

or $\quad 4(n^2 + 1) = (2n + 1)(2n - 1) + 5$.

Since $2n + 1 \mid n^2 + 1$, we also have $2n + 1 \mid 4(n^2 + 1)$.

Thus $\quad 2n + 1 \mid (2n + 1)(2n - 1) + 5$,

implying $2n + 1 \mid 5$.

This means that $\quad 2n + 1 = 1 \quad$ or $\quad 2n + 1 = 5$,

giving $\quad n = 0 \quad$ or $\quad n = 2$.

To complete the proof of Trigg's result, it remains to show that there is essentially only one way to arrange the numbers 1, 2, 3, . . ., 19 in a 2-ring array so that it is magical. (Two arrangements are considered the same if one can be obtained from the other by a rotation or a reflection.) By some neat analysis, Allaire was able to settle the matter by having a computer investigate 70 simple cases (simple for the computer). Counting compiling time, it took the computer about 17 seconds to find (and establish the uniqueness of) what Clifford W. Adams had taken 47 years to find. We conclude this topic with a sample or two of Allaire's further analysis.

Let the numbers in the cells be denoted $x_1, x_2, \ldots, x_{19}$, as shown. In each of the "principal" directions, there are $2n + 1 = 5$ "lines" of constant sum s. Thus the sum of all the cells is $1 + 2 + \cdots + 19 = 5s$, giving $s = 38$.

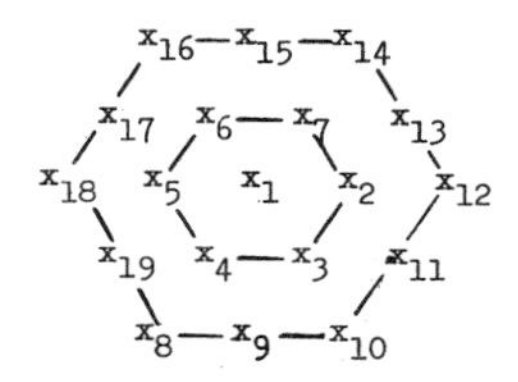

We see first of all that the numbers 1 and 2 cannot go in an outer corner cell. We have, for the case of x_{10},

$$s = x_8 + x_9 + x_{10} = x_{10} + x_{11} + x_{12} = 38.$$

Thus

$$x_8 + x_9 + 2x_{10} + x_{11} + x_{12} = 76,$$

or

$$x_8 + x_9 + x_{11} + x_{12} = 76 - 2x_{10}.$$

Now, if x_{10} is either 1 or 2, we have $76 - 2x_{10} \geqq 72$. But the maximum value of $x_8 + x_9 + x_{11} + x_{12}$ is only $19 + 18 + 17 + 16 = 70$.

It is almost as easy to see that the middle cell of an outer edge is equal to the sum of the central cell and the two opposite cells of the inner ring, e.g., $x_9 = x_1 + x_6 + x_7$.

Adding along the five lines indicated by the sketch

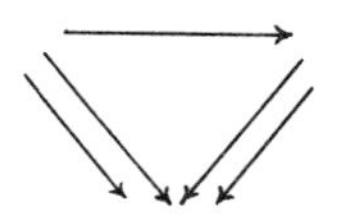

we count x_9 twice, and every other x_i once, except x_1, x_6, x_7, which are omitted altogether. However, 5 lines give a sum equal to the sum of all the x_i. Thus, taking x_9 a second time must just compensate for omitting x_1, x_6, and x_7. Hence $x_9 = x_1 + x_6 + x_7$. As a corollary, we see that the middle cell of an outer edge, being the sum of three x_i, must be at least $1 + 2 + 3 = 6$. We conclude, then, that the numbers 1 and 2 cannot occur in the outer ring at all.

Finally, we show that the central cell must contain a number x_1 which is no greater than 8. Adding along the eight lines indicated by

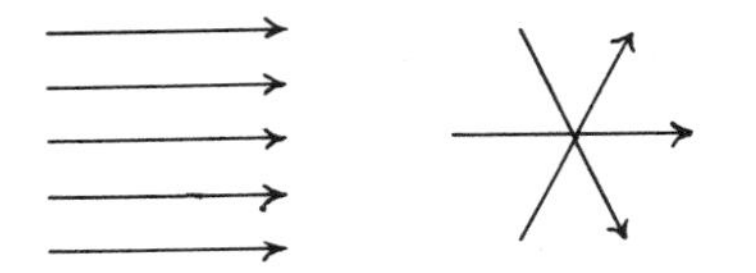

we obtain x_1 four times, each cell in the inner ring twice, each outer corner twice, and each cell in the middle of an outer ring once, that is,

$$4x_1 + 2(\text{inner ring}) + 2(\text{outer corners}) + (\text{outer mid-edges}) = 8(38).$$

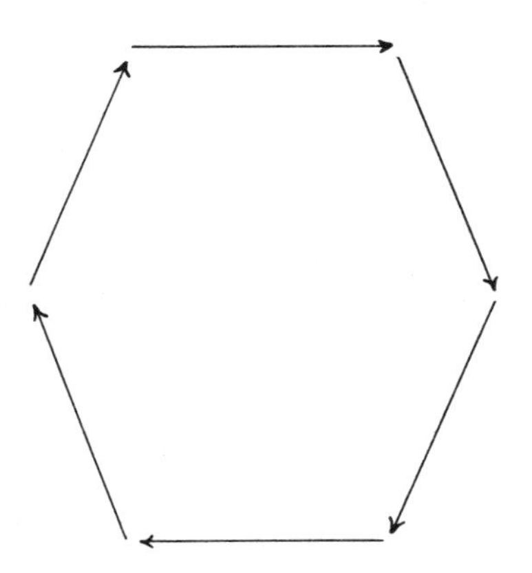

However, the six lines around the outside give 2(outer corners) + (outer mid-edges) = 6(38). Subtracting gives

$$4x_1 + 2(\text{inner ring}) = 2(38), \text{ and } 2x_1 + \text{inner ring} = 38.$$

Hence $2x_1 = 38 - (x_2 + x_3 + x_4 + x_5 + x_6 + x_7)$. The maximum value here is $38 - (1 + 2 + 3 + 4 + 5 + 6) = 17$. Thus

$$2x_1 \leqq 17, \text{ implying } x_1 \leqq 8.$$

POSTSCRIPT. It has just recently come to light that Clifford W. Adams was not the first person to consider and to find the magic hexagon. It seems that one Martin Kühl of Hanover, Germany, made the discovery around 1940, but his work was not published. Also, the hexagon appears without comment under the name of Tom Vickers in the Mathematical Gazette, December 1958, page 291. Charles W. Trigg's report on Clifford W. Adams' work, containing his proof of the uniqueness of the hexagon, is contained in the Recreational Mathematics Magazine, January 1964.

Exercises

1. Determine whether or not it is possible in chess for a knight, on an ordinary 8×8 board, to go from one corner square to the opposite corner square and to light exactly once on each of the other squares in the process.

2. A classroom has 5 rows of 5 desks per row. The teacher requests each pupil to change his seat by going either to the seat in front, the one behind, the one to his left, or to the one on his right (of course, not all these options are possible for all students). Determine whether or not this directive can be carried out.

3. Prove that the number of people at the opera next Thursday who will shake hands an odd number of times is an even number.

4. Let n denote an odd integer greater than 1. Let A be an $n \times n$ symmetric matrix such that each row and each column consists of some permutation of the integers $1, 2, 3, \ldots, n$. Show that each of the integers $1, 2, 3, \ldots, n$ must occur on the main diagonal of A.

5. It is desired to invert an entire set of n upright cups by a series of moves in each of which any selected $n - 1$ cups are turned over. Show that this can always be done if n is even, but never if n is odd.

6. A 6×6 chessboard is covered by 18 2×1 dominoes. Show that, no matter how this is done, it is possible to cut along one of the horizontal or down one of the vertical rulings of the board without cutting through a domino.

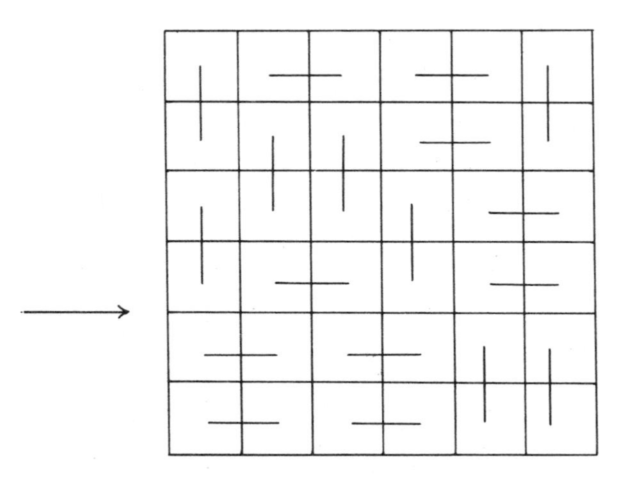

7. What is the smallest number of bishops which can be arranged on an ordinary 8×8 chessboard so that every square is controlled by at least one bishop? (Assume a bishop controls the square it occupies.)

8. In how many ways can a king in chess move from one corner square to the diagonally opposite square if he always moves towards his goal? (Of course, not all moves are "directly" towards the goal.)

9. In a 3×3 magic square on the numbers 1 to 9, show that the sum of the numbers in any row, column, or diagonal is 15, that the number in the center is 5, and that 1 cannot go in a corner.

10. A $3''$-cube can be cut into 27 $1''$-cubes by six cuts, two parallel to each pair of opposite faces at a distance of $1''$ from each of the opposite faces. During this operation it is not necessary to move any of the pieces produced. Is it possible to reduce the number of cuts if one is free to rearrange the pieces between cuts?

11. Thirteen bricks of dimensions $1 \times 1 \times 2$ contain only 26

cubic units of volume, one unit less than that of a $3 \times 3 \times 3$ cube. Is it possible to arrange thirteen such bricks to form a $3 \times 3 \times 3$ cube with the central $1 \times 1 \times 1$ cube missing?

References and Further Reading

1. Fryer and Berman, An Introduction to Combinatorics, Academic Press, New York, 1972.

2. Yaglom and Yaglom, Challenging Mathematical Problems with Elementary Solutions, Holden-Day, San Francisco, 1964.

3. I. Niven, Mathematics of Choice, Random House New Mathematics Library Series, Vol. 15.

4. C. L. Liu, Introduction to Combinatorial Mathematics, McGraw-Hill, New York, 1968.

CHAPTER **7**

THE KOZYREV-GRINBERG THEORY OF HAMILTONIAN CIRCUITS

1. Introduction. Hamiltonian circuits were introduced in the essay on Louis Pósa. (We recall that a Hamiltonian circuit in a graph is one which passes through every vertex exactly once.) As noted there, no necessary and sufficient conditions are known for a graph to possess a Hamiltonian circuit.

Any graph can be represented by a diagram, consisting of points for vertices and arcs for edges, which lies in a plane. In many cases, however, there is no way to put in the edges without having some of them intersect, that is, cross over one another at points which are not vertices. For example, there is no way to embed in a plane a complete graph on five vertices (i.e., 5 vertices with all

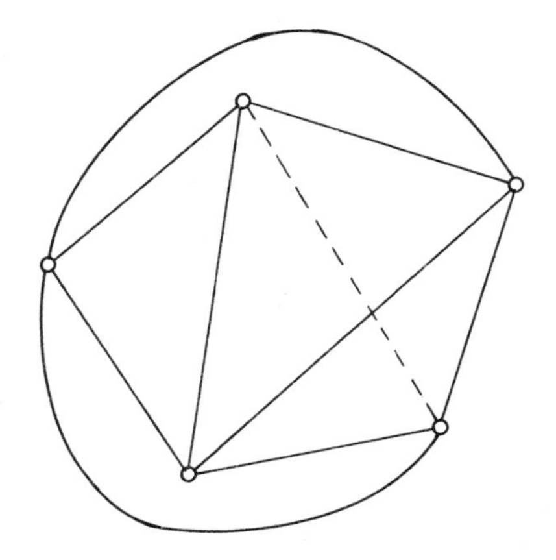

possible edges) without some pair of edges crossing. Graphs which can be represented in a plane in such a way that no edges meet at a point other than a vertex are called **planar** graphs. These constitute an important class of graphs, and many interesting discoveries have been made concerning them.

In 1968, a necessary condition for a planar graph to have a Hamiltonian circuit was presented at a conference on Graph Theory at Manebach, Germany. This result was communicated to the meeting by the German mathematician Sachs under the names of two Soviet mathematicians, Kozyrev and Grinberg. I have recently been informed that it was Grinberg who made the mathematical discovery, with Kozyrev helping to make it known. Whatever the correct dispensation of credit might be, the work is now firmly associated with both men and we continue to use both names when referring to it. Their elegant approach is so simple and straightforward that it is remarkable no one thought of it until now. Just how far-reaching and important their work is remains to be seen.

2. The Kozyrev-Grinberg Theory. We begin by considering a planar graph G, with n vertices, which does have a Hamiltonian

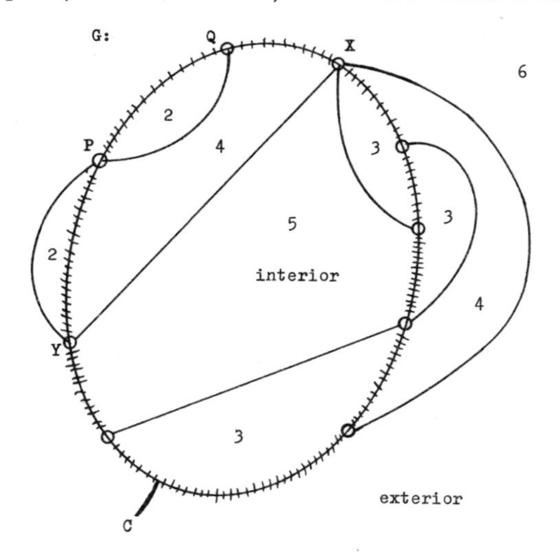

circuit C. The edges which do not belong to the circuit occur as "diagonals" in the interior or in the exterior of C. Whether a diagonal is put inside or outside C is often an optional matter. It makes no difference to us how one chooses to distribute the edges. Let us suppose that all the options have been exercised and that we have before us some definite planar representation of the graph G.

Consider first the interior of the circuit C. Suppose that d diagonals occur there. Since G is a planar graph, none of its edges intersect. Thus a diagonal splits into two parts the region through which it passes. Thinking of putting in the diagonals one at a time, we see that the insertion of a diagonal increases by one the number of regions inside the circuit. Consequently the d diagonals divide the interior of C into $d + 1$ regions.

Let f_i denote the number of these regions in C which are bounded by i edges, that is, the number of i-gons in C. Here i might be as small as 2, in the case of a di-gon (like PQ), and as great as n, if there happen to be no diagonals at all inside C. By the above result we have

$$\sum_{i=2}^{n} f_i = f_2 + f_3 + \cdots + f_n = d + 1.$$

In each i-gon inside C let us write the number i, denoting the number of edges in its boundary. The sum of all these numbers in the regions would give the total number of edges bounding the regions, counting each diagonal twice, since it bounds two of the regions, and counting once each of the n edges in the circuit C (each bounds only one region). Accordingly, the grand total would be $2d + n$. However, it is clear that for each $i = 2, 3, \ldots, n$ the number i would occur in the sum f_i times, since there are f_i i-gons. Thus we have

$$\sum_{i=2}^{n} if_i = 2f_2 + 3f_3 + \cdots + nf_n = 2d + n.$$

From the earlier result, we have $d = [\sum_{i=2}^{n} f_i] - 1$. Substituting for

d, we obtain

$$\sum_{i=2}^{n} if_i = 2[\sum_{i=2}^{n} f_i] - 2 + n = [\sum_{i=2}^{n} 2f_i] - 2 + n,$$

giving

$$\sum_{i=2}^{n} (i-2)f_i = n - 2.$$

Turning to the exterior of C, we obtain in identical fashion that

$$\sum_{i=2}^{n} (i-2)f_i' = n - 2,$$

where f_i' denotes the number of i-gons in the exterior of C. In deriving this result, we again make use of the fact that d diagonals provide $d+1$ regions. Thus among these regions we do need to include the infinite part of the exterior. Accordingly, for example, one diagonal gives two regions, one of infinite extent. The bounding edges of this infinite region are all the edges around the outside of the entire graph G, whether they be diagonals or edges of the circuit C. Combining our two main results, we obtain by subtraction the Kozyrev-Grinberg necessary condition

$$\sum_{i=2}^{n} (i-2)(f_i - f_i') = 0.$$

The values of f_i and f_i' depend on the course taken through the graph by the circuit C. If C is the unknown object of one's investigations, the value of this condition would appear to be lost in the circularity of the situation. However, the equation involves i as well as f_i and f_i', and this opening is all we need in order to put the thing to good use.

3. Three Simple Applications.

(a) The given graph Q (see the diagram) does possess various Hamiltonian circuits. Show, however, that any such circuit which contains one of the edges A, B, must avoid the other.

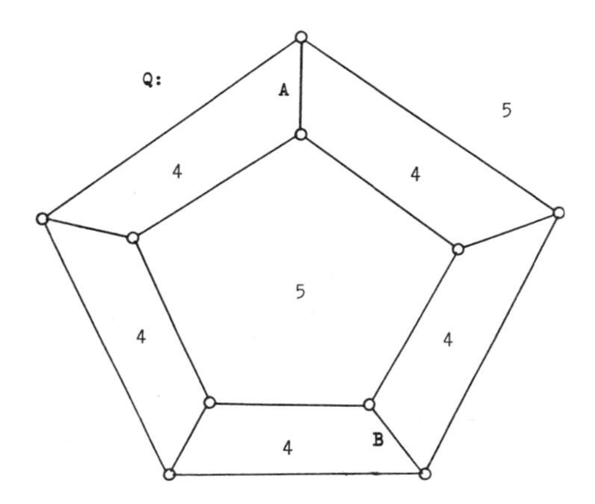

The faces of Q provide only 5 quadrilaterals and 2 pentagons. Any Hamiltonian circuit of Q must split these faces between its interior and exterior in such numbers that

$$2(f_4 - f_4') + 3(f_5 - f_5') = 0.$$

This implies that $f_4 - f_4'$ is divisible by 3. Since there are only five quadrilaterals altogether, the only possible values for f_4 and f_4' are 4 and 1, making $f_4 - f_4'$ either 3 or -3.

Now each of the edges A and B separates a pair of 4-gons. Thus a Hamiltonian circuit which contains the edge A would have one of A's quadrilaterals inside and the other outside. Similarly, a Hamiltonian circuit containing the edge B would split B's quadrilaterals. For both A and B to belong to a Hamiltonian circuit, at least two of the five quadrilaterals must be inside and at least two of them outside the circuit. This makes impossible the four-one split for f_4 and f_4' which is necessary to satisfy the Kozyrev-Grinberg condition. Consequently, no Hamiltonian circuit of Q can contain both the edges A and B.

A slightly simpler solution was noticed by Frank Allaire, now a graduate student at the University of Waterloo. He employs the often useful device of adding vertices on the edges of a graph under investigation. Let Q' denote the graph obtained from Q by adding a vertex M on A and a vertex N on B. Now any Hamiltonian cir-

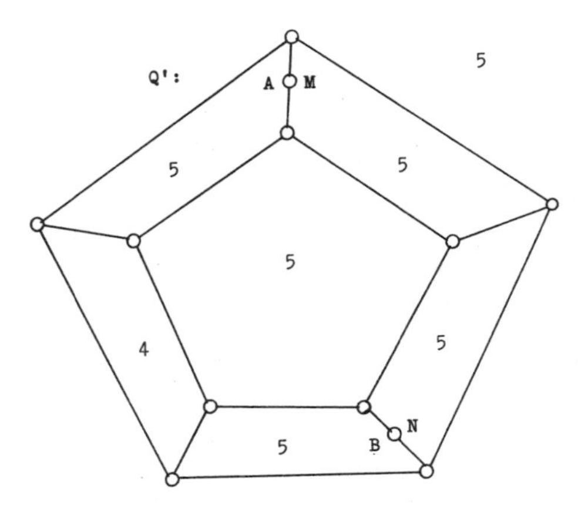

cuit of Q which contains both A and B determines a Hamiltonian circuit of Q'. However, it follows directly from the Kozyrev-Grinberg relation that Q' has no Hamiltonian circuits. A Hamiltonian circuit of Q' would again necessitate

$$2(f_4 - f_4') + 3(f_5 - f_5') = 0,$$

implying that $f_4 - f_4'$ is divisible by 3. However, with the extra vertices, Q' has only one 4-gon, making $f_4 - f_4'$ either 1 or -1. (QED)

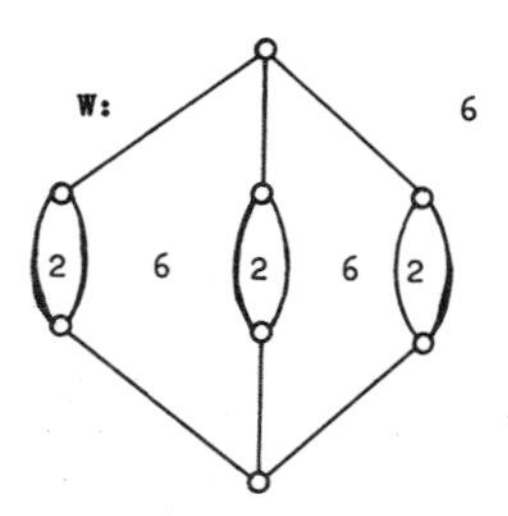

(b) Next we show that the given graph W (in the diagram) has no Hamiltonian circuit.

The six faces consist of 3 di-gons and 3 hexagons. The Kozyrev-

Grinberg condition gives

$$0(f_2 - f_2') + 4(f_6 - f_6') = 0,$$

implying that f_6 and f_6' are equal. This, however, is impossible, since $f_6 + f_6' = 3$.

(c) Thirdly, we show that the given graph G (in the diagram) does not possess a Hamiltonian circuit.

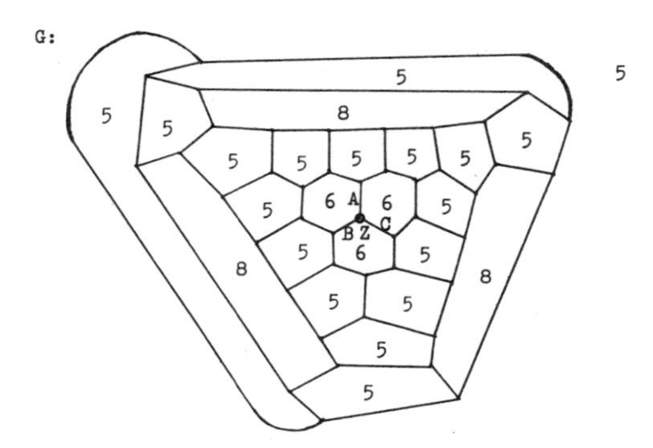

We observe that the faces have either 5, 6, or 8 edges. Thus a Hamiltonian circuit must yield

$$3(f_5 - f_5') + 4(f_6 - f_6') + 6(f_8 - f_8') = 0.$$

Hence $f_6 - f_6'$ must be divisible by 3. But there are only 3 6-gons. Thus $f_6 - f_6'$ must equal 3 or -3. That is to say, any Hamiltonian circuit of G must put all three hexagons inside or all three outside in order to make one of f_6, f_6' 3 and the other 0. Accordingly, the circuit cannot contain any of the edges A, B, C at vertex Z because each lies between two hexagons. Thus Z is an inaccessible vertex, making a Hamiltonian circuit impossible.

4. Tait's Conjecture. In 1880 the English mathematician P. G. Tait conjectured that every member of a special class of planar graphs, the so-called 3-connected cubic graphs, possesses a Hamiltonian circuit. For the purposes of the present discussion it is not necessary to go into the precise meaning of 3-connected (cubic simply means that each vertex has valence 3). Tait's conjecture was of vital importance to everyone who worked on the famous four-color problem, for its affirmation would prove that all plane maps can be colored with four colors. The activity surrounding this monumental unsolved problem has been so great that the Norwegian mathematician Oystein Ore has declared that, in the last hundred years, almost every advance in graph theory can be traced to the desire to solve it. Thus the negation of Tait's conjecture in 1946 by the brilliant effort of the Anglo-Canadian mathematician W. T. Tutte was of widespread interest. Tutte's graph T, pictured below, is an example of a 3-connected, cubic planar graph which does not have a Hamiltonian circuit.

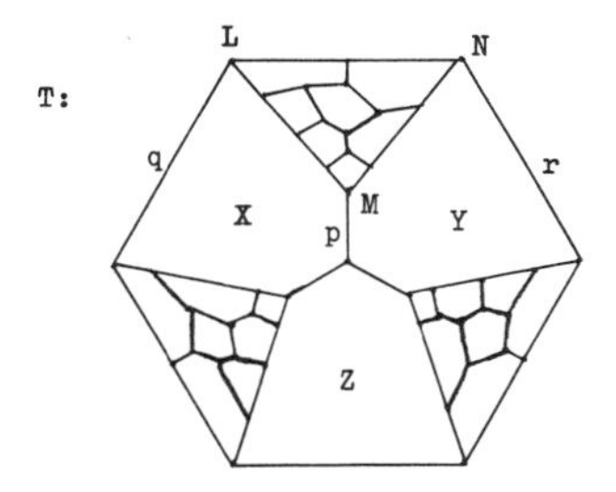

Tutte's over-riding accomplishment here is the construction of the graph. Once in possession of it, it is not too difficult to check it out and show that it is non-Hamiltonian. However, at first brush, it does not seem that the Kozyrev-Grinberg theory is going to help. It was in the summer of 1972 that Wayne Watts, a high school mathematics teacher from metropolitan Toronto, successfully applied the Kozyrev-Grinberg theory to prove that Tutte's graph is non-Hamiltonian. I might add that Tutte was very pleased to see this neat piece of work. Watts' argument follows.

We begin with a few general observations. Each edge of a graph separates a pair of adjacent faces. If an edge belongs to a circuit, then the faces it separates lie on opposite sides of the circuit, and conversely. If an edge does not belong to a circuit, then both its faces lie on the same side of the circuit, and conversely.

Now we proceed indirectly; that is, let us assume that Tutte's graph T does possess a Hamiltonian circuit C, and then go on to use the Kozyrev-Grinberg condition to deduce a contradiction. Referring to the diagram of T, consider the three decagons X, Y, Z. Suppose some two of them, say X and Y, were to lie outside a Hamiltonian circuit C. Then their common edge p could not be an edge of C. Likewise, edges q and r cannot belong to C because the infinite outer face also lies outside C. Thus the triangular section LMN has no edges of C connecting it to the rest of the graph. This defies the existence of C, giving the conclusion that not more than one of the decagons X, Y, Z can lie outside C. Equivalently, at least two of the decagons, say X and Y, must lie inside C.

In this case, too, their common edge p cannot belong to C. By similar arguments, both the edges q and r would have to belong to C (the infinite face is outside). Consequently, the circuit must enter and leave the triangular section LMN by the edges q and r. Accordingly, in the section LMN, itself, there must exist a (Hamiltonian) *path* linking L and N which passes through every vertex of LMN exactly once.

Now we focus our attention on the triangular section LMN, stripped of the rest of T and its connecting edges p, q, and r. The Hamiltonian path in LMN is made into a Hamiltonian *circuit* by introducing an edge LN in the exterior of LMN. Now the Kozyrev-Grinberg theory applies, giving the equation

$$1(f_3 - f_3') + 2(f_4 - f_4') + 3(f_5 - f_5') + 6(f_8 - f_8') = 0.$$

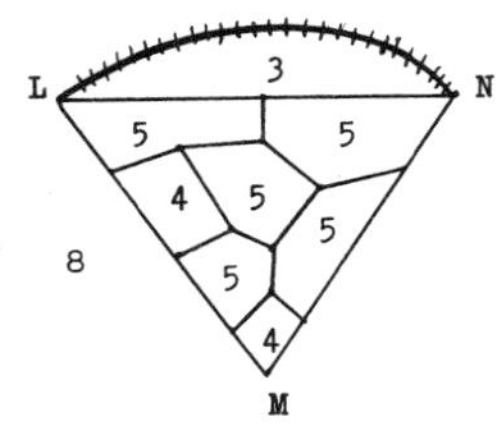

Since the edge LN does belong to the Hamiltonian circuit in LMN, the single 3-gon face it bounds must occur inside the circuit, implying that $f_3 - f_3' = 1 - 0 = 1$. Also, the infinite 8-gon face lies outside the circuit, giving $f_8 - f_8' = 0 - 1 = -1$. Thus our equation reduces to

$$1 + 2(f_4 - f_4') + 3(f_5 - f_5') - 6 = 0,$$

or

$$2(f_4 - f_4') + 3(f_5 - f_5') = 5. \tag{1}$$

At the vertex M, there are only two edges available for our Hamiltonian circuit; thus both must belong to the circuit. Consequently, the 4-gon face at M lies inside the circuit. There is only one other 4-gon face. Thus either both 4-gons lie inside the circuit or there is one on each side, that is, $f_4 - f_4'$ is either $2 - 0 = 2$ or $1 - 1 = 0$. If it has the value 2, then equation (1) implies that 3 divides 1; if the value 0, (1) implies that 3 divides 5. In either case we have a contradiction.

We conclude this chapter by showing that the graph G, devised

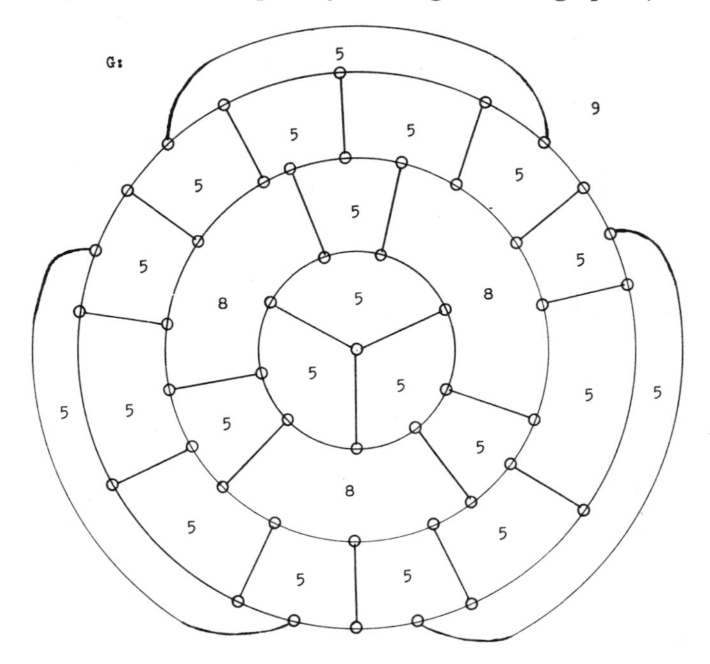

by Kozyrev and Grinberg, is a simpler counterexample to Tait's conjecture. All its faces have either 5, 8, or 9 edges. In order for G to have a Hamiltonian circuit, it must be that $3(f_5 - f_5') + 6(f_8 - f_8') + 7(f_9 - f_9') = 0$. But there is only one 9-gon face, making $7(f_9 - f_9')$ either 7 or -7. However, the equation implies that this term must be divisible by 3. (QED)

Exercises

1. Show that any Hamiltonian circuit in the graph G which contains the edge x must also contain the edge y.

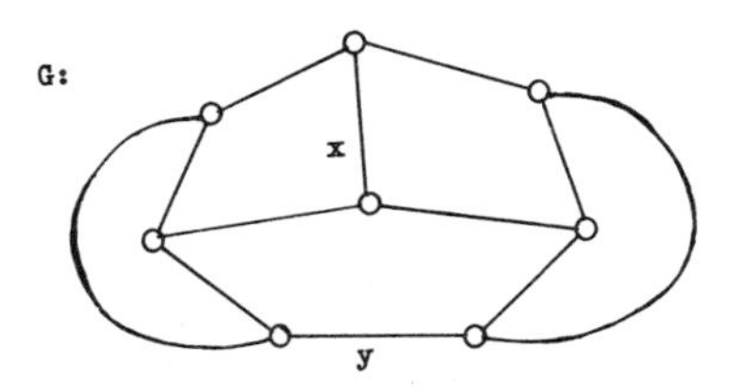

2. Show that any Hamiltonian circuit in the graph G must contain exactly two of the edges x, y, z. Show that a Hamiltonian circuit which contains both edges p and q cannot also contain x.

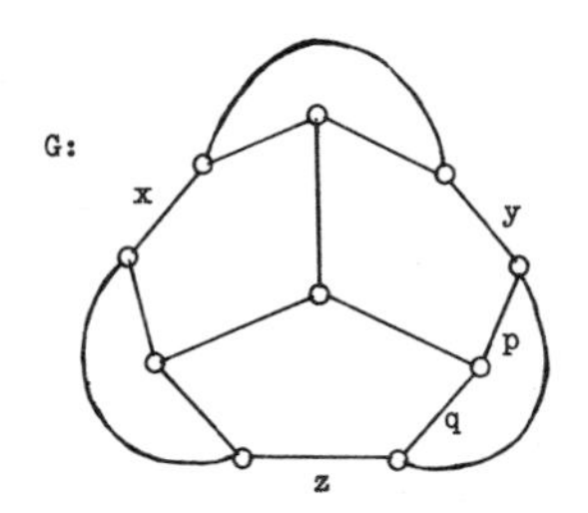

3. Show that the pentagon P must lie outside any Hamiltonian circuit in the graph G, and that a Hamiltonian circuit in G must contain exactly 4 of the edges of P.

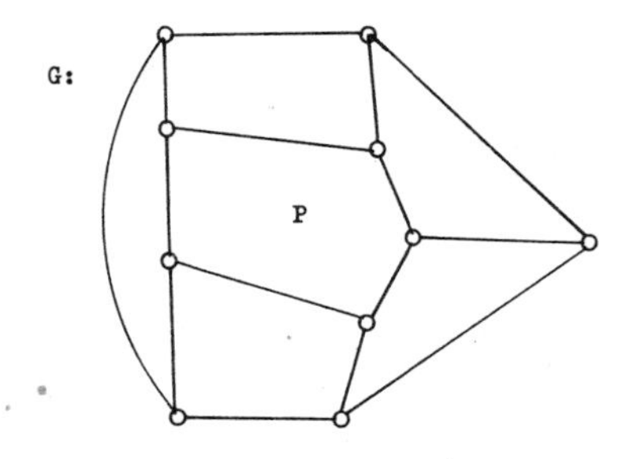

4. Use the Kozyrev-Grinberg relation to show that none of the graphs below possesses a Hamiltonian circuit.

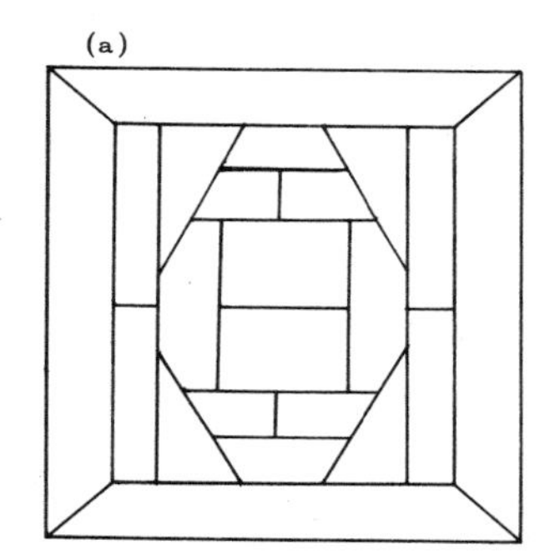

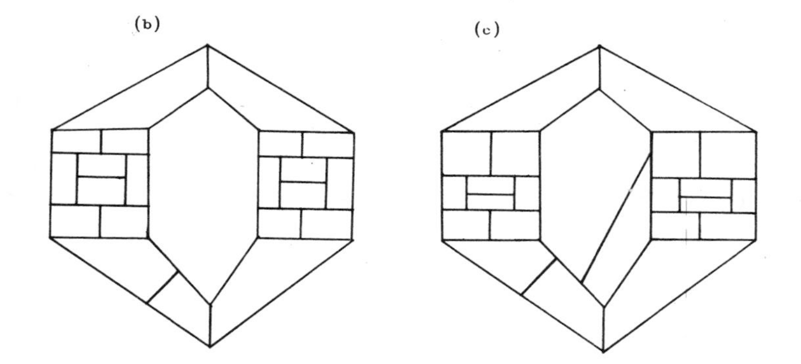

The graphs in parts (b) and (c) were obtained in 1971 by G. B. Faulkner at the University of Waterloo.

5. Pick up Watts' argument at the point of considering section *LMN* with extra edge *LM* added to complete a Hamiltonian circuit. Complete the proof by adding vertices on edges, as in Allaire's solution of problem 3(a), and using the Kozyrev-Grinberg condition.

CHAPTER **8**

MORLEY'S THEOREM

From ancient times mathematicians have shown particular interest in performing geometric constructions with straightedge and compass. The impossibility of trisecting a general angle with these instruments has tended to direct attention away from problems involving the trisectors of angles. This helps account for the late appearance of the following delightful theorem:

The adjacent pairs of the trisectors of the angles of a triangle always meet at the vertices of an equilateral triangle.

This was discovered only in 1904 by the Anglo-American geometer Frank Morley (1860–1937).

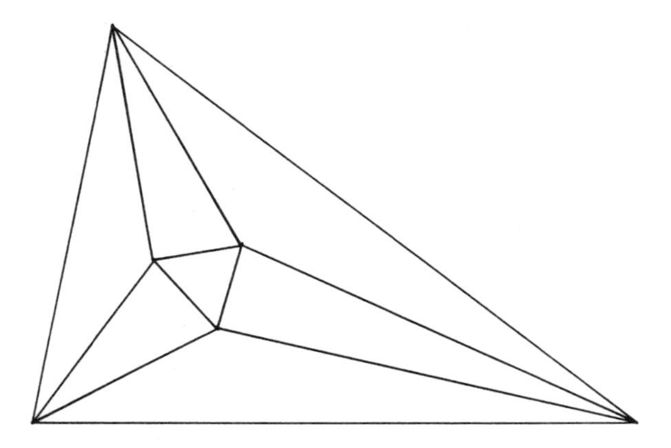

Morley was a remarkable person. Although he spent the last 50 years of his life in the United States (mostly at Johns Hopkins University), he never gave up his British citizenship. Besides being a first-rate mathematician, he excelled at chess, having the distinction of once beating Emmanuel Lasker when Lasker was the reigning champion of the world.

Morley came upon the theorem "from above." In 1924 he disclosed its method of discovery. Even without understanding all the details, it is illuminating to have a glimpse into Morley's "home territory". He found that:

"If a variable cardioid touch the sides of a triangle the locus of its center, that is, the center of the circle on which the equal circles roll, is a set of 9 lines which are 3 by 3 parallel, the directions being those of the sides of an equilateral triangle. The meets of these lines correspond to double tangents; they are also the meets of certain pairs of trisectors of the angles, internal and external, of the first triangle."

Fortunately such a striking result was bound to move others to attempt a simpler proof. In 1909 a very appealing proof was supplied by M. T. Naraniengar, the simplicity of which, to this day, has never been surpassed at the elementary level of exposition. His proof was rediscovered in 1922 by J. M. Child.

NARANIENGAR'S PROOF.

Let the given triangle ABC have angles $A = 3a$, $B = 3b$, $C = 3c$. Let the trisectors adjacent to BC meet at P, and let the other trisectors of angles B and C meet at S. Then PB and PC bisect angles in ΔSBC. Accordingly, P is the incenter of this triangle, and PS bisects angle S.

On each side of SP construct, at P, angles of 30 degrees to yield points Q and R on SC and SB. This makes triangles PRS and PQS congruent, showing that $PR = PQ$. Since $\angle QPR = 60$ degrees, the triangle PQR is equilateral. It remains to show, then, that AR and AQ trisect angle A.

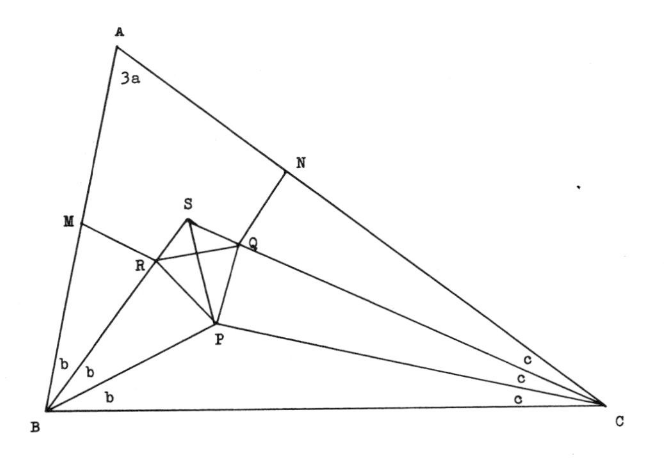

Mark off on BA the length BP to give M, and on CA the length CP to give N. Then the triangles MBR and BRP are congruent, giving $MR = RP$. Similarly $QN = PQ$. Since ΔPQR is equilateral we obtain

$$MR = RQ = QN.$$

Next we compute the size of angle MRQ. For ΔPRS we note that exterior angle BRP equals the sum of the interior angles at S and P, that is, $\angle BRP = \frac{1}{2}S + 30°$. And angle BRM is the same size. Thus we have

$$\angle MRQ = 360° - 2(\tfrac{1}{2}S + 30°) - 60° = 240° - S.$$

An identical argument shows that $\angle RQN$ is also $240° - S$. Since the angles in a triangle add up to 180 degrees, we have

$$S = 180° - (2b + 2c) = 180° - \tfrac{2}{3}(3a + 3b + 3c) + 2a$$
$$= 180° - 120° + 2a = 60° + 2a.$$

Thus

$$\angle MRQ = \angle RQN = 240° - (60° + 2a) = 180° - 2a.$$

Now we turn to the intuitively obvious fact that the equal segments MR, RQ, and QN, which are equally inclined to one another, have their endpoints on a circle. By constructing the perpendicular bisectors XO, YO of MR and RQ to give O, it is an easy exercise

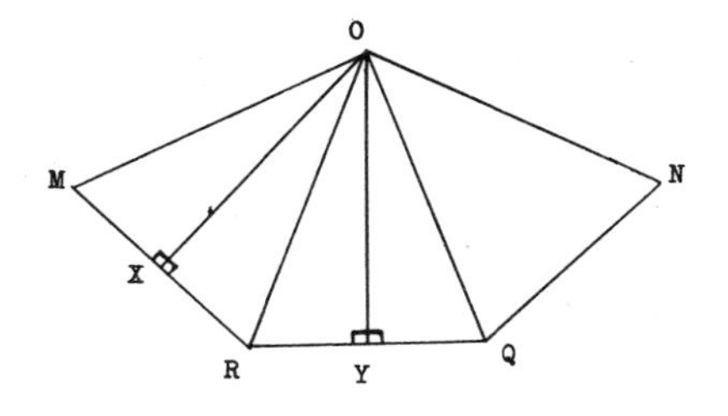

to prove that the three triangles OMR, ORQ, OQN are congruent. This makes $ON = OQ$, implying that the circle with center O and radius MO passes through all four points M, R, Q, N. Also, the radii OR and OQ bisect the angles at R and Q, and the lines XO, YO bisect angles at O. We have

$$\angle MRO = \tfrac{1}{2}\angle MRQ = \tfrac{1}{2}(180^\circ - 2a) = 90^\circ - a.$$

This gives $\angle XOR = a$, $\angle MOR = 2a$, and leads to $\angle MON = 6a$.

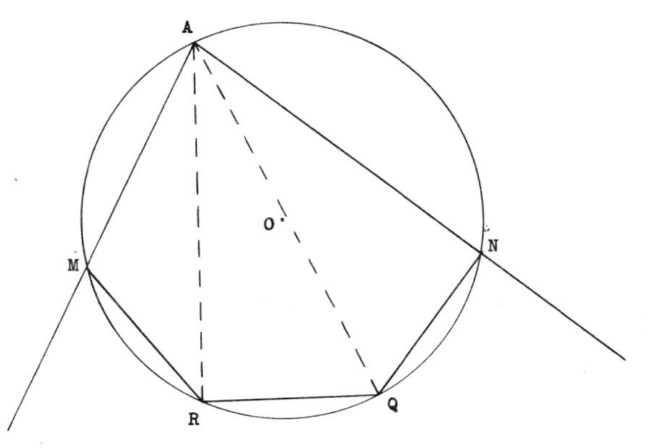

In this circle, then, the chord MN subtends an angle of $6a$ at the center O. Accordingly, at the circumference, MN subtends an angle of $3a$. But MN subtends an angle of $3a$ at the vertex A of the given ΔABC. Thus the circle must go through A. Since MR, RQ, and QN subtend $2a$ at the center O, each subtends an angle a at A on the circumference. Hence AR and AQ trisect angle A. (QED)

Exercises

1. The incenter of a triangle is usually described as the point of intersection of the bisectors of the angles. Show that the incenter I of ΔABC can also be specified as the point along the bisector of $\angle A$ at which the opposite side BC subtends an angle of $90° + A/2$.

In Naraniengar's proof, compute the angles BPR and CPQ. It is by specifying these angles, instead of constructing the 30-degree angles at P, that J. M. Child obtained the points Q and R.

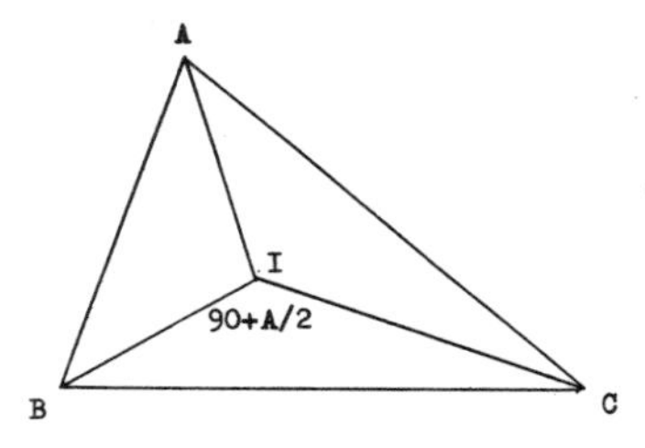

2. Begin with equilateral triangle PQR. On its sides construct outwardly three isosceles triangles to yield points P', Q', R', by constructing equal angles x at Q and R, equal angles y at P and R, equal angles z at P and Q.

If x, y, and z are chosen so that each is less than 60 degrees and their sum $x + y + z = 120$ degrees, use Euclid's parallel postulate to show that $Q'R$ and $R'Q$ must meet at some point X. (Similarly,

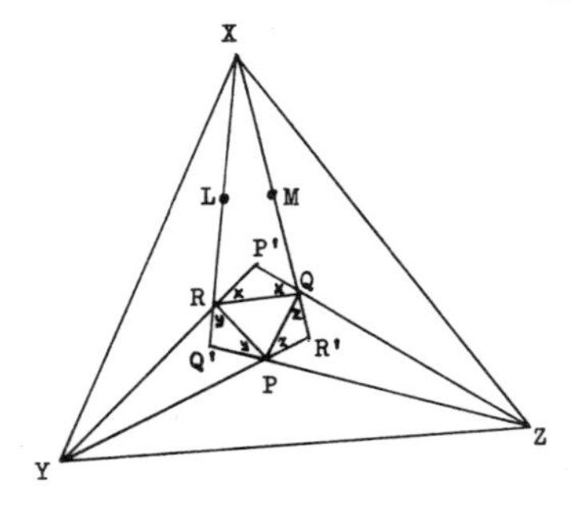

then, we also obtain Y and Z.) If $x = 60° - A/3$, $y = 60° - B/3$, $z = 60° - C/3$, prove that the angles in ΔXYZ are A, B, and C. Then Morley's theorem for ΔABC follows immediately from a dilatation which carries ΔXYZ into congruency with ΔABC. (See reference [4].)

3. Using the notation of exercise 2, prove that PP', QQ', RR' are concurrent.

4. In 1914 (Edinburgh Math. Soc. Proceedings, p. 120), W. E. Philip published a proof of Morley's theorem which proceeded according to the following plan. Work through his proof, computing the angles and verifying his argument.

Let the given triangle ABC have angles $A = 3a$, $B = 3b$, $C = 3c$. At least one of these angles is acute; let A be acute. Let the trisectors of angles B and C meet at D and L, as shown. Then D is the incenter of ΔBCL. Let the incircle touch sides at Q and R, as shown. Let DQ and DR be produced to give H and K on AB, AC, respectively. Then congruent triangles give $QH = DQ$, $DR = RK$, all four segments being equal to the radius of the incircle. Let tangent KP be drawn and extended to give F on BL.

Then: (i) ΔDKP is a 30°-60°-90° triangle. (ii) $\angle QDR = 120° - 2a$. (iii) $\angle QHF = 30° - a$. (iv) $\angle DHK = 30° + a$. (v) $\angle FHK = 2a$. (vi) $\angle HKF = a$. (vii) $\angle HFK = 180° - 3a = 180° - A$. This makes A, H, F, K concyclic. And in this circle the chord HF

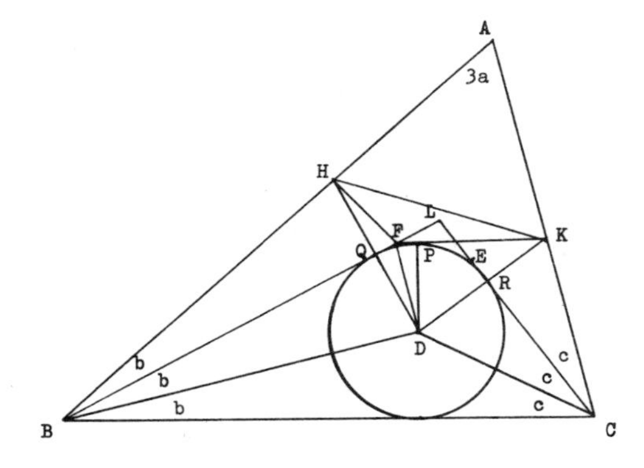

subtends at the circumference an angle $HKF = a$. Thus $\angle HAF = a$, making AF a trisector of angle A.

Similarly, the tangent to the incircle of ΔBCL from H gives a point E on CL such that EA is the other trisector of angle A. The triangle DHK and the circle center D form quite a symmetrical figure; the triangle is isosceles ($DH = DK$) and the center D of the circle is the vertex common to the equal sides. Without repeating the argument, then, we see that $\angle EDR = \angle FDQ = 30° - a$, and it also follows that $FD = DE$. Thus show that $\angle FDE = 60°$, from which we conclude ΔDEF is equilateral, proving the theorem.

5. Show that Morley's Theorem holds also in the case of the trisection of the exterior angles of a triangle.

References and Further Reading

1. Coxeter and Greitzer, Geometry Revisited, Random House New Mathematical Library Series, Vol. 19.

2. M. T. Naraniengar, Mathematical Questions and Their Solutions From the Educational Times (New Series), Vol. 15, 1909, p. 47.

3. J. M. Child, A proof of Morley's theorem, Math. Gaz. (1922) 171.

4. H. S. M. Coxeter, Introduction to Geometry, Wiley, New York, 1961.

5. American Mathematical Society, Semi-Centennial Publications, vol. 1—History.

CHAPTER **9**

A PROBLEM IN COMBINATORICS

The diagonals of a square divide the interior into four regions. The diagonals of a regular pentagon divide its interior into eleven regions. In this essay we shall work out the number of regions into which a polygon of n sides is divided by its diagonals.

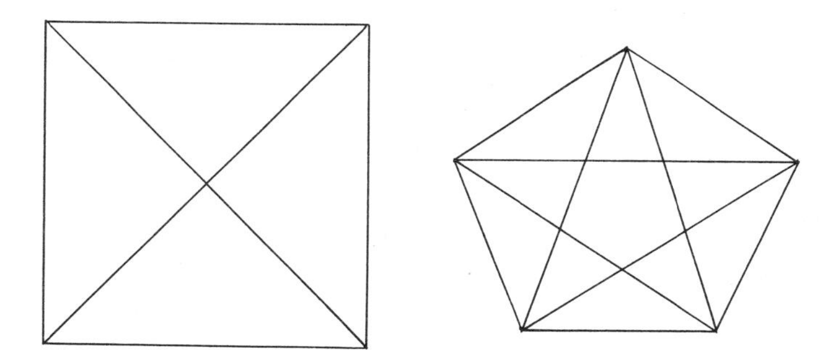

In order to have every diagonal entirely inside the figure, we shall deal only with convex polygons. Now, the shifting of vertices around the plane would move the diagonals and change the sizes and shapes of the regions inside. If three or more diagonals are moved to pass through the same point, a region of the interior would collapse to the vanishing point, thus changing the number of regions. In order to obtain the maximum case, we consider only n-gons for which no three diagonals are concurrent. Our problem, then, is the following:

Into how many regions do the diagonals of a convex n-gon divide the interior if no three are concurrent inside the n-gon?

The matter of counting things is often quite a sophisticated art, containing ideas of great ingenuity. While we shall not consider all known methods, we indicate the richness of this fast-growing subject of combinatorics by giving three elegant solutions to our problem.

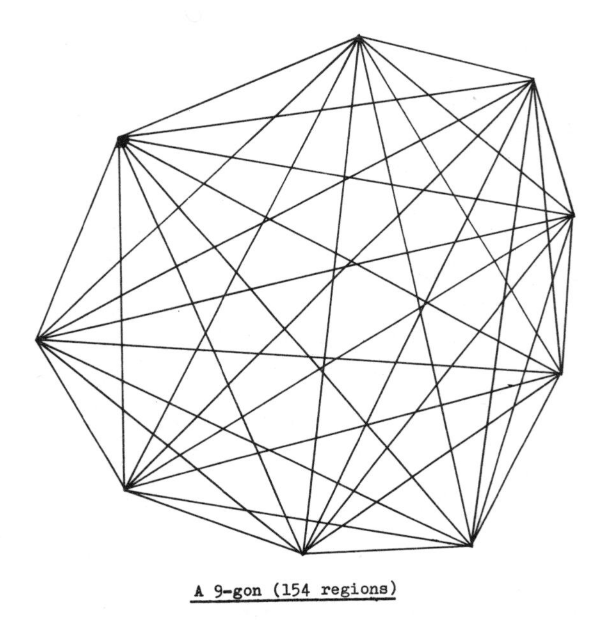

A 9-gon (154 regions)

1. Counting Vertices and Angles. Let n_k denote the number of k-gons among the regions in our n-gon. Suppose that m is the maximum number of sides in any polygon of the decomposition. Now, counting each triangle at three vertices, each quadrilateral at four vertices, etc., the total number of vertices altogether, including a lot of duplication, is

$$3n_3 + 4n_4 + 5n_5 + \cdots + mn_m.$$

Each interior point, being the intersection of two diagonals, is a vertex of four of the regions in the n-gon. In this total, then, each interior point is counted four times. Each vertex A_v of the given n-gon is a vertex of $n - 2$ triangles cut off by the diagonal $A_{v-1}A_{v+1}$, and so is counted $n - 2$ times. Thus we have

$$3n_3 + 4n_4 + \cdots + mn_m$$
$$= 4(\text{number of interior points}) + (n - 2)(\text{number of vertices}).$$

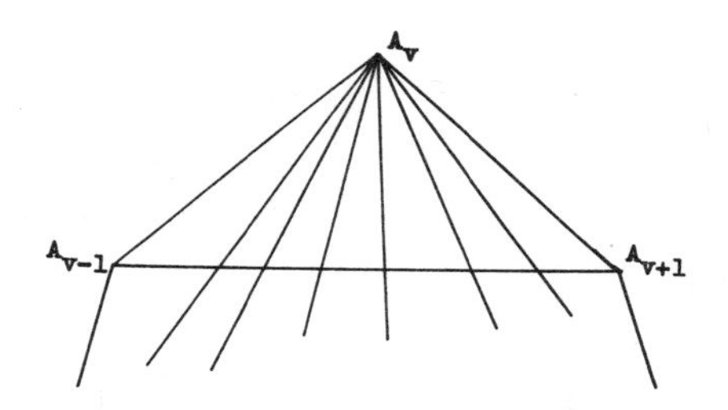

Because there is a 1-1 correspondence between the interior points and the sets of four vertices which can be selected from the vertices of the given n-gon, we see that the number of interior points is simply $\binom{n}{4}$, the number of such quadruples of vertices. Consequently, we have the result

$$(1) \qquad 3n_3 + 4n_4 + \cdots + mn_m = 4\binom{n}{4} + (n - 2)n.$$

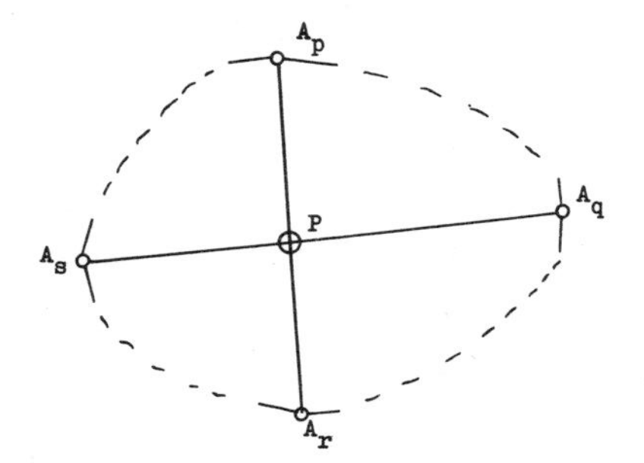

Next we count up all the angles in all the regions. The sum of the angles of a k-gon is $(k - 2)180$ degrees. Thus the total for all regions is

$$n_3 \cdot 180° + n_4 \cdot 360° + n_5 \cdot 540° + \cdots + n_m \cdot (m - 2)180°.$$

This time there is no duplication at all. Now we are free to add these angles in any order we wish. The four angles at an interior point add up to 360 degrees. The sum of all the interior angles, then, is just $360\binom{n}{4}$ degrees. At the vertices of the given n-gon there is an additional $(n - 2)180$ degrees. Altogether we have

$$n_3 \cdot 180 + n_4 \cdot 360 + \cdots + n_m \cdot (m - 2)180 = 360\binom{n}{4} + (n - 2)180.$$

Dividing through by 180, we obtain

$$(2) \quad n_3 + 2n_4 + 3n_5 + \cdots + (m - 2)n_m = 2\binom{n}{4} + n - 2.$$

Subtracting this from our previous equation (1) gives

$$2n_3 + 2n_4 + 2n_5 + \cdots + 2n_m = 2\binom{n}{4} + (n - 1)(n - 2).$$

Dividing by 2, we obtain the desired result:

$$\text{Number of regions} = n_3 + n_4 + n_5 + \cdots + n_m = \binom{n}{4} + \frac{(n-1)(n-2)}{2} = \binom{n}{4} + \binom{n-1}{2}$$

2. Euler's Formula. In this method we use the well-known formula of Euler for simple polyhedra and plane maps, $V - E + F = 2$, where V, E, F denote the number of vertices, edges, and faces. The F in this formula includes 1 for the infinite outer face of the figure. Since we want only the number of interior faces, we have

$$F = (\text{number of interior faces}) + 1.$$

Substituting, we obtain

$$\text{number of interior faces} = 1 + E - V.$$

The value of V is simply the sum of the number of interior points and the number of vertices of the given n-gon, that is,

$$V = \binom{n}{4} + n.$$

The whole solution turns on finding E.

We can distinguish three kinds of edges in the figure—(a) sides of the given n-gon, (b) both endpoints interior points, (c) one end interior and the other a vertex of the given n-gon. We see that the number of edges of type (c) is $n(n - 3)$ because $n - 3$ of them occur at each of the n vertices of the given n-gon. Now at each interior point we have four edges meeting. Thus the number $4\binom{n}{4}$ counts all the interior edges; in fact it duplicates every edge of type (b) while counting the edges of type (c) once. We may denote this by the equation

$$4\binom{n}{4} = 2(b) + (c).$$

In the same notation we have $2(a) + (c) = 2n + n(n - 3)$. Adding these results gives

$$4\binom{n}{4} + 2n + n(n - 3) = 2(a) + 2(b) + 2(c).$$

But the right side of this equation is just $2E$. Hence

$$E = 2\binom{n}{4} + n + \frac{n(n - 3)}{2}.$$

Accordingly, we obtain:

Number of interior faces

$$= 1 + \left[2\binom{n}{4} + n + \frac{n(n-3)}{2} \right] - \left[\binom{n}{4} + n \right]$$

$$= \binom{n}{4} + \frac{n(n-3)}{2} + 1$$

$$= \binom{n}{4} + \binom{n-1}{2}, \text{ as before.}$$

3. A Reduction Process. I have saved the neatest solution until last. In this method we simply tear down the figure by removing diagonals one at a time, noting as we go the decrease in the number of regions.

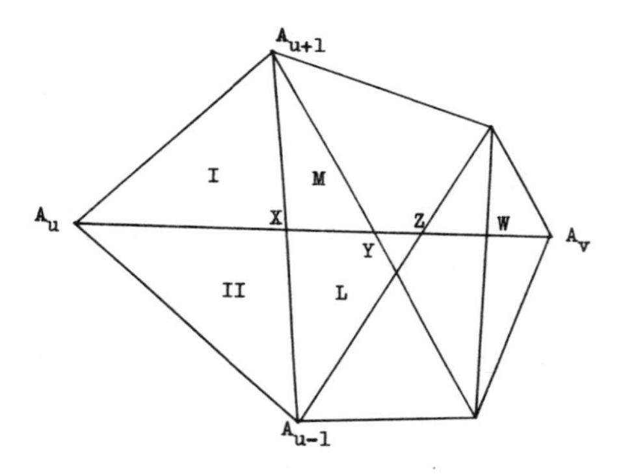

Referring to the diagram, consider the number of regions which are lost by the removal of diagonal A_uA_v. Imagine this diagonal to be peeled off like a strip of adhesive tape. When the section A_uX has been lifted, the regions I and II, which it had separated, now go together to form a single region. As the section XY is taken up, regions L and M join to form a single region. Similarly for every section down the diagonal. Thus the number of regions lost is equal to the number of sections along the diagonal. This number is one more than the number of points of intersection, $X, Y, Z, \ldots,$ which the other diagonals produce on A_uA_v.

Since each point of intersection has only two diagonals through it, the removal of one of the diagonals causes the point of intersection to vanish from the other diagonal, too. Thus, by the time a diagonal is removed, many or even all of its points of intersection may have been stripped from it already. Consequently, the decrease in the number of regions is one more than the number of points of intersection which are left on the diagonal at the time it is removed. Thus in keeping track of the decreases throughout the entire process of removing the diagonals, each point of intersection figures once and only once. Therefore the total number of regions lost in this process is

[(number of intersections on first diagonal) + 1]
+ [(number of intersections left on second diagonal) + 1]
+ [(number of intersections left on third diagonal) + 1]
+ . . . .
+ [(number of intersections left on last diagonal) + 1].

In this sum there is a term for every diagonal, and there is a 1 in every term. Therefore the value of this sum is the number of interior points of intersection plus the number of diagonals. We observe that of the $\binom{n}{2}$ possible segments joining a pair of the n given vertices, n are sides of the polygon. Thus the number of diagonals is $\binom{n}{2} - n$, giving us the result:

$$\text{Total number of regions lost} = \binom{n}{4} + \left[\binom{n}{2} - n\right].$$

However, after removing all the diagonals, there is still one region left inside the n-gon (the interior, itself). Consequently, the number of regions which must have been present at the beginning is

$$\left[\binom{n}{4}+\binom{n}{2}-n\right]+1=\binom{n}{4}+\frac{n(n-1)}{2}-(n-1)$$
$$=\binom{n}{4}+(n-1)\frac{(n-2)}{2}$$
$$=\binom{n}{4}+\binom{n-1}{2}.$$

Exercises

1. Inside a convex n-gon m points are taken at random. Line segments are drawn to join pairs of the $m + n$ points in the figure (the m points and the n vertices) in any way such that no two of the segments intersect and all the regions in the figure are triangles. How many triangles are there?

2. Prove that every simple polyhedron has at least two faces with the same number of edges.

3. N straight lines in general position divide the plane into various parts. How many of these regions are infinite?

4. Determine the total number of squares of all sizes and positions on an ordinary 8×8 chessboard. Also compute the total number of rectangles.

5. Prove that it is always possible to partition a given obtuse-angled triangle into seven or fewer acute-angled triangles.

6. Show how to divide a square into 8 acute-angled triangles.

7. Into how many solid regions is 3-dimensional space divided by n planes, assuming that every three planes meet at exactly one point and that no four do?

8. If all the faces of a simple polyhedron have central symmetry prove that at least six of the faces are parallelograms.

9. At each vertex of a simple polyhedron exactly three faces

meet. If each face is either a pentagon or a hexagon, how many pentagonal faces are there?

References and Further Reading

1. Yaglom and Yaglom, Challenging Mathematical Problems with Elementary Solutions, Holden-Day, San Francisco, 1964.

2. Fryer and Berman, Introduction to Combinatorics, Academic Press, New York, 1972.

CHAPTER **10**

MULTIPLY-PERFECT, SUPERABUNDANT, AND PRACTICAL NUMBERS

1. Introduction. In ancient times the Pythagoreans classified the natural numbers as deficient, perfect, and abundant according to the sum of their divisors. The number 6 was called **perfect**, because the sum of its proper divisors, 1, 2, and 3, is 6, itself. The numbers 8 and 12, respectively, are **deficient** and **abundant** because the proper divisors of 8, namely 1, 2, and 4, total only 7, and those of 12, which are 1, 2, 3, 4, and 6, add up to more than 12. It turns out that there are many deficient and abundant numbers, but that perfect numbers are very rare. By 1972 only 24 perfect numbers have been discovered. The first four are

$$6, 28, 496, 8128.$$

The fifth is in the millions, and then they start to get big. In the writings of Euclid there occurs a formula for even perfect numbers:

$m = 2^{n-1}(2^n - 1)$ is perfect if the factor $2^n - 1$ is a prime number.

In the eighteenth century, Euler proved the converse, that every even perfect number is the product of two numbers, 2^{n-1} and $2^n - 1$, where $2^n - 1$ is a prime. With this formula, one can show easily that every even perfect number must end in either a 6 or an 8, and that when it ends in an 8, it ends in 28. Surprisingly, no odd perfect number has ever been found, and it is an open question whether or not one exists.

In number theory there is a well-known function $\sigma(n)$ which denotes the sum of all the divisors of the natural number n. This includes the divisor n, itself. Thus, in terms of $\sigma(n)$, the perfect numbers n are those for which $\sigma(n) = 2n$. The deficient and abundant numbers are defined similarly.

We conclude the introduction with the derivation and application of a formula which express $\sigma(n)$ in terms of the prime decomposition of n:

$$n = p_1^{a_1} p_2^{a_2} \cdots p_k^{a_k}.$$

While the algebraic notation forces the derivation to appear complicated, the ideas involved place no strain on us whatsoever. The divisors d of n are the numbers

$$d = p_1^{b_1} p_2^{b_2} \cdots p_k^{b_k}$$

where each b_i is restricted to the range $0 \leqq b_i \leqq a_i$ (d will not divide n if any b_i exceeds its a_i). But these numbers are precisely the terms in the expansion of the product

$$(1 + p_1 + p_1^2 + \cdots + p_1^{a_1})(1 + p_2 + p_2^2 + \cdots + p_2^{a_2}) \cdots (1 + p_k + p_k^2 + \cdots + p_k^{a_k}).$$

This, then, is the value of $\sigma(n)$. Since each bracket is a geometric progression, we have

$$\sigma(n) = \frac{p_1^{a_1+1} - 1}{p_1 - 1} \cdot \frac{p_2^{a_2+1} - 1}{p_2 - 1} \cdots \frac{p_k^{a_k+1} - 1}{p_k - 1}.$$

Now we are able to prove the useful result that, if m and n are relatively prime,

$$\sigma(mn) = \sigma(m) \cdot \sigma(n).$$

Suppose the prime decompositions of m and n are

$$m = p_1^{a_1} p_2^{a_2} \cdots p_k^{a_k} \quad \text{and} \quad n = q_1^{b_1} q_2^{b_2} \cdots q_t^{b_t}.$$

Since m and n are relatively prime, no p_i and q_j denote the same prime number. Thus

$$mn = p_1^{a_1} p_2^{a_2} \cdots p_k^{a_k} q_1^{b_1} q_2^{b_2} \cdots q_t^{b_t},$$

and we have

$$\sigma(mn) = \frac{p_1^{a_1+1} - 1}{p_1 - 1} \cdot \frac{p_2^{a_2+1} - 1}{p_2 - 1} \cdots \frac{p_k^{a_k+1} - 1}{p_k - 1} \cdot \frac{q_1^{b_1+1} - 1}{q_1 - 1} \cdots \frac{q_t^{b_t+1} - 1}{q_t - 1},$$

which is clearly $\sigma(m) \cdot \sigma(n)$.

2. Multiply-Perfect Numbers. In recent times, the notion of perfect number has been generalized as follows:

If, for the natural number n, the value of $\sigma(n) = kn$, where k is a natural number, then n is said to be k-tuply perfect.

Accordingly, the ancient perfect numbers correspond to $k = 2$. It is easily verified that the numbers 120 and 672 are triply-perfect, since $\sigma(120) = 360$ and $\sigma(672) = 3(672)$. We use the notation "p_k number" for a k-tuply perfect number. Thus

$$n = 2178540 = 2^2 \cdot 3^2 \cdot 5 \cdot 7^2 \cdot 13 \cdot 19$$

is a p_4 number, since, for it, $\sigma(n) = 4n$. Even with all the attention these numbers have attracted down through the years, the following two basic questions have never been answered:

(i) Is there an infinity of p_k numbers, counting those for all $k = 2, 3, 4, \ldots$?

(ii) Is any p_k number odd?

We consider now three little problems concerning p_k numbers which have been solved, and delightfully so.

(a) If n is a p_3 number and is not a multiple of 3, prove that $3n$ is a p_4 number.

Since n is a p_3 number, we have $\sigma(n) = 3n$. Because n is not a multiple of 3, we see that 3 and n are relatively prime, giving $\sigma(3n) = \sigma(3) \cdot \sigma(n)$. But $\sigma(3) = 1 + 3 = 4$, yielding $\sigma(3n) = 4 \cdot \sigma(n)$. Since $\sigma(n) = 3n$, we have $\sigma(3n) = 4(3n)$, showing $3n$ to be a p_4 number.

(b) If $3n$ is a p_{4k} number and n is not divisible by 3, show that n is a p_{3k} number.

We have $\sigma(3n) = \sigma(3)\cdot\sigma(n) = 4\cdot\sigma(n)$, since 3 and n are relatively prime. But $\sigma(3n) = 4k(3n)$ since $3n$ is a p_{4k} number. Thus $4\cdot\sigma(n) = 4k(3n)$, giving $\sigma(n) = 3k(n)$, implying n is a p_{3k} number.

(c) If n is a p_3 number and 3 divides n but 5 and 9 do not divide n, show that $45n$ is a p_4 number.

We have $\sigma(n) = 3n$. Now $n = 3k$ for some natural number k, where 3 does not divide k since 9 does not divide n. Also, then, $\sigma(n) = \sigma(3k) = \sigma(3)\cdot\sigma(k) = 4\cdot\sigma(k)$. Now $\sigma(45n) = \sigma(3^3\cdot 5\cdot k) = \sigma(3^3)\cdot\sigma(5)\cdot\sigma(k) = 40\cdot 6\cdot\sigma(k) = 60[4\cdot\sigma(k)]$, since 3^3, 5, and k are relatively prime. But $4\sigma(k) = \sigma(n)$. Hence

$$\sigma(45n) = 60\cdot\sigma(n) = 60(3n) = 180n = 4(45n),$$

implying that $45n$ is a p_4 number.

We conclude this section with a final application of the formula for $\sigma(n)$.

THEOREM. *Every p_5 number has more than 5 different prime divisors.*

Let the prime decomposition of the p_5 number n be denoted by $n = p_1^{a_1}p_2^{a_2}\cdots p_k^{a_k}$. Then $\sigma(n) = 5n$ gives the equation

$$\frac{p_1^{a_1+1}-1}{p_1-1}\cdot\frac{p_2^{a_2+1}-1}{p_2-1}\cdots\frac{p_k^{a_k+1}-1}{p_k-1} = 5p_1^{a_1}p_2^{a_2}\cdots p_k^{a_k}.$$

Dividing through by n, we obtain

$$\frac{p_1-(p_1^{a_1})^{-1}}{p_1-1}\cdot\frac{p_2-(p_2^{a_2})^{-1}}{p_2-1}\cdots\frac{p_k-(p_k^{a_k})^{-1}}{p_k-1} = 5.$$

Neglecting the negative terms in the numerators, we obtain the inequality

$$\frac{p_1}{p_1-1}\cdot\frac{p_2}{p_2-1}\cdots\frac{p_k}{p_k-1} > 5.$$

Now the values of these factors, $p_i/(p_i-1)$, are terms of the strictly decreasing sequence 2/1, 3/2, 5/4, 7/6, 11/10, . . ., obtained by considering the prime numbers in increasing order. Thus the product of any five of these terms is greatest when they are the first five terms. And the product of fewer than five terms is less than the product of the first five terms. But the product of even the first five terms is only 77/16, which is <5. Thus, in order to satisfy the above inequality, more than five terms must be used, implying that n has more than five distinct prime divisors.

3. Superabundant Numbers. In 1944, the mathematicians Erdös and Alaoglu defined superabundant numbers as follows:

A natural number n is superabundant if and only if

$$\sigma(n)/n \quad \text{exceeds} \quad \sigma(k)/k \qquad \text{for all } k < n.$$

The values of $\sigma(n)$ for $n = 1, 2, 3, 4, 5$ are, respectively, 1, 3, 4, 7, and 6, giving the values for $\sigma(n)/n$ to be 1, 3/2, 4/3, 7/4, and 6/5. Thus we see that 2 and 4 are superabundant and that 3 and 5 are not. We consider now a nice proof of the fact that there exists an infinity of superabundant numbers.

Let us denote the divisors of the natural number n, in increasing order, by $d_1 = 1, d_2, d_3, \ldots, d_t = n$. Then $n/d_1, n/d_2, \ldots, n/d_t$ is the same set of divisors, in the reverse order. Their sum is the same, giving

$$\sigma(n) = \sum_{d_i \mid n} \frac{n}{d_i}.$$

Dividing by n, we obtain

$$\frac{\sigma(n)}{n} = \sum_{d_i \mid n} \frac{1}{d_i},$$

the sum of the reciprocals of the divisors of n.

Now as n runs through the natural numbers 1, 2, 3, . . ., it takes on the values of all factorials, $m!$. For $n = m!$, the divisors of n include all the numbers $1, 2, 3, \ldots, m$, and likely many more, giving

$$\frac{\sigma(n)}{n} = \sum_{d_i|n} \frac{1}{d_i} \geqq \frac{1}{1} + \frac{1}{2} + \frac{1}{3} + \cdots + \frac{1}{m}.$$

But this partial sum of the famous harmonic series grows beyond all bounds as m increases without limit. That is to say, the values of $\sigma(n)/n$ grow beyond all bounds as n runs through the natural numbers. If we denote the number $\sigma(n)/n$ by u_n, then the sequence $u_1, u_2, u_3, \ldots$ has no upper bound.

We observe that if the term u_n of this sequence exceeds all preceding terms u_k, then n is a superabundant number. Because the sequence has no upper bound, it can never stop producing greater and greater terms u_n (not necessarily consecutive terms) which do exceed all previous terms. The subscripts n associated with this endless parade of high-points constitute an infinity of superabundant numbers.

4. Practical Numbers. In 1948, A. K. Srinivasan defined a practical number as follows:

A natural number n is practical if and only if, for all $k \leqq n$, k is the sum of distinct proper divisors of n.

All even perfect numbers are practical. In fact, whether or not the number $2^n - 1$ is a prime number, the number $m = 2^{n-1}(2^n - 1)$ is practical for all $n = 2, 3, 4, \ldots$.

Proof. The proper divisors of m include the two groups of numbers

(A) $$1, 2, 2^2, 2^3, \ldots, 2^{n-1},$$

and

(B) $$2^n - 1, 2(2^n - 1), 2^2(2^n - 1), \ldots, 2^{n-2}(2^n - 1).$$

If $2^n - 1$ happens to be composite, there are other divisors, too.

However, using just these proper divisors, we can add suitable ones to produce all k less than or equal to m.

We consider three cases:

(i) *If* $k \leqq 2^{n-1}$, then k is the sum of numbers from the set (A). Expressing k in the scale of 2 shows which of these numbers add up to it.

(ii) *If* $k = m$, then k is the sum of all the numbers in (A) and (B). This is easily verified by adding the two geometric series involved.

(iii) *If* $2^{n-1} < k < m$, then k can be expressed, upon division by $2^n - 1$, as $k = (2^n - 1)t + r$, where $0 \leqq r < 2^n - 1$. We observe that unless the quotient t is $< 2^{n-1}$, the value of k is at least m. Consequently, the number t, itself, is the sum of numbers from the set (A) (think binary again). In fact, we do not even need to use the number 2^{n-1}, since it is greater than t.

Considering binary notations, we see that the numbers in the set (A) are capable of yielding all numbers up to $2^n - 1$. Thus r, too, is the sum of numbers from (A). As a result,

$$k = (2^n - 1)(2^i + 2^j + \cdots + 2^q) + (2^t + 2^u + \cdots + 2^v)$$

where the greatest power in the first large sum is $2^q \leqq 2^{n-2}$ and the greatest power in the second is $2^v \leqq 2^{n-1}$. Expanding this expression for k, then, shows it to be the sum of terms from (B) followed by terms from (A). (QED)

5. Related Numbers.

(a) QUASIPERFECT NUMBERS. Quasiperfect numbers are the least abundant, that is, $\sigma(n)$ exceeds $2n$ by the least amount, namely 1. For quasiperfect numbers, then, we have $\sigma(n) = 2n + 1$. Alternatively, quasiperfect numbers are natural numbers which are the sum of their nontrivial divisors, i.e., the divisors excluding both 1 and n. This gives $n = \sigma(n) - 1 - n$, as above. Unfortunately, no quasiperfect numbers are known.

Closely related to these numbers are the least deficient numbers, that is, the natural numbers n for which $\sigma(n) = 2n - 1$. It is easily seen that powers of 2 fall into this category.

(b) SEMIPERFECT NUMBERS. A natural number n is semiperfect if it is the sum of some set of distinct proper divisors of n.

For example: for $n = 12$, we have $12 = 6 + 4 + 2$. Clearly a semiperfect number cannot be deficient, and every perfect number is semiperfect. It appears that most of the abundant numbers are semiperfect. Exceptions seem so rare that abundant numbers which are not semiperfect have been dubbed "weird" numbers. The three smallest weird numbers are 70, 836, and 4030. No odd weird number has yet been found, and Paul Erdös has offered $10 for the first one and $25 for the first proof that none exists. It is known, however, that an infinity of weird numbers exists. The study of weird numbers was initiated by Stan Benkoski of Pennsylvania State University, and a forthcoming joint paper with Erdös shows that the sequence of weird numbers has positive Schnirelmann density (the Schnirelmann density of a sequence of natural numbers is the greatest lower bound of the fractions $A(n)/n$, where $A(n)$ denotes the number of terms of the sequence $\leqq n$. The density of a sequence lacking the number 1, then, is automatically zero. In order to avoid this undesirable feature, the number 1 is usually added to sequences which do not originally have it. This must be done for weird numbers. The density of the odd numbers is 1/2, while that of the squares, primes, and powers of 2 is zero.) This indicates that the sieve of weird numbers across the natural numbers is somewhat "thicker" than the primes and the other sequences of density zero.

We close this essay with the remark that there seems to be no end to the fascinating properties of the natural numbers. There are also other exotic numbers which we have not touched upon, such as "powerful numbers".

Exercises

1. Prove that every multiple mn, $m \geqq 2$, of a perfect or abundant number n is abundant.

2. Prove that every divisor of a deficient number is deficient.

3. Prove that p^n, p a prime, is deficient.

4. Prove that $\sigma(n)$ is odd if and only if n is a square or twice a square.

5. Prove the theorem of Euler: *An odd perfect number must be of the form* $m = p^{4a+1} \cdot Q^2$, *where* p *is an odd prime which is relatively prime to* Q.

Reference and Further Reading

1. W. Sierpiński, Elementary Theory of Numbers, Warszawa (1964).

CHAPTER **11**

CIRCLES, SQUARES, AND LATTICE POINTS

1. Introduction. As we saw in the chapter on the Orchard Problem, the lattice points (x, y) of a coordinate plane have both x and y integers. They occur regularly in rows and columns at the vertices of unit squares. It is obvious that a small circle can be drawn around any of these points so as to exclude all other lattice points. With a little experimenting one finds that it is not too difficult to determine circles which contain exactly 2, exactly 3, and exactly 4 lattice points in their interiors.

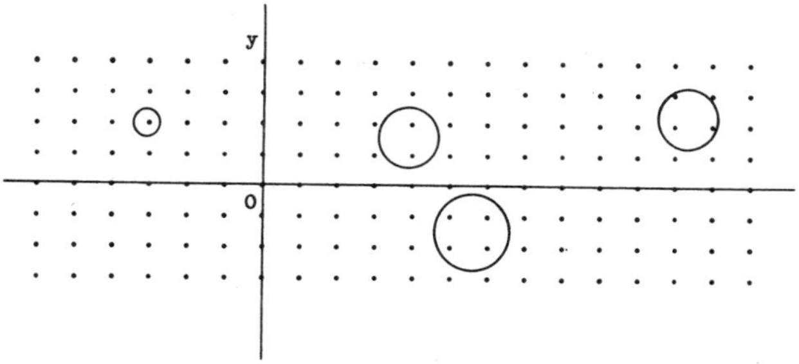

It is not always easy to find a circle which contains a specified large number of lattice points inside it. We begin this essay by establishing the theorem that, for every natural number n, there exists in the plane a circle which contains in its interior exactly n lattice points.

The theorem follows easily from the fact that no two lattice points are the same distance from the point $(\sqrt{2}, 1/3)$. We prove this shortly. Because of this, the lattice points can be ordered in a sequence $p_1, p_2, p_3, \ldots$, according to their distance from $(\sqrt{2}, 1/3)$; p_1 is closest, p_2 next closest, and so on. Then the circle with center $(\sqrt{2}, 1/3)$ which passes through p_{n+1} contains in its interior precisely the n lattice points $p_1, p_2, \ldots, p_n$.

We complete the proof by proceeding indirectly. Suppose that the two lattice points (a, b) and (c, d) are the same distance from the point $(\sqrt{2}, 1/3)$. Then

$$(a - \sqrt{2})^2 + (b - 1/3)^2 = (c - \sqrt{2})^2 + (d - 1/3)^2.$$

Separating rational and irrational parts, we obtain

$$2(c - a)\sqrt{2} = c^2 + d^2 - a^2 - b^2 + \tfrac{2}{3}(b - d).$$

The left-hand side is irrational or zero while the right-hand side is rational. For equality, both sides must be zero. Thus

$$c = a, \quad \text{and} \quad c^2 + d^2 - a^2 - b^2 + \tfrac{2}{3}(b - d) = 0.$$

Since $c = a$, the latter gives $d^2 - b^2 - \frac{2}{3}(d - b) = 0$ and

$$(d - b)(d + b - \tfrac{2}{3}) = 0.$$

Since d and b are integers, the factor $d + b - \frac{2}{3}$ cannot be zero. Thus we must have

$$d - b = 0, \text{ giving } d = b.$$

Since $c = a$, the two lattice points (a, b) and (c, d) must be the same point; contradiction. (QED)

Hugo Steinhaus has even proved that, for every natural number n, there exists a circle of *area* n which contains in its interior exactly n lattice points.

2. Schinzel's Theorem. We turn now to circles which have exactly n lattice points on the circumference. The cases $n = 1$ and $n = 2$ are trivial. However, even the early case $n = 3$ gives pause for thought. In 1958 André Schinzel, a Polish mathematician from Warsaw, published a delightful proof that, for every natural number n, there exists in the plane a circle which has exactly n lattice points on its circumference. His proof has the added feature that it identifies the circle by its equation. His argument proceeds as follows.

The cases of n even and n odd are considered separately. Schinzel shows that

(i) for $n = 2k$, the circle with center $(1/2, 0)$ and radius $\frac{1}{2}\cdot 5^{(k-1)/2}$ passes through exactly n lattice points, and that

(ii) for $n = 2k + 1$, the circle with center $(1/3, 0)$ and radius $\frac{1}{3}\cdot 5^k$ passes through exactly n lattice points.

He uses the well-known formula from number theory that the number $r(n)$ of integral solutions (x, y) to the equation $x^2 + y^2 = n$ is four times the difference between the number of divisors of n of the form $4k + 1$ and the number of the form $4k + 3$. This is denoted by

$$r(n) = 4(d_1 - d_3).$$

The derivation of this formula belongs to elementary number theory, but it is too long to be included here. We take the liberty of using it without proof, and we note that it counts (x, y) and (y, x) as two different solutions. Thus, for example, $r(1) = 4(1 - 0) = 4$, the solutions being $(1, 0)$, $(0, 1)$, $(-1, 0)$, $(0, -1)$.

When n is an even number $2k$, we consider the integral solutions of the equation $x^2 + y^2 = 5^{k-1}$. All the divisors of 5^{k-1} are powers of 5. Thus every divisor is of the form $4k + 1$. Since there are k such powers of 5 up to 5^{k-1}, the number of integral solutions is $4(k - 0) = 4k$. These solutions go together to form $2k$ pairs of reversals, (x, y) and (y, x). Now 5^{k-1} is odd, implying that one of x and y is odd and the other is even. Thus exactly one solution in each of the $2k$ pairs has first term odd and second term even. We use this result later.

For each lattice point (p, q) that occurs on the circle with center $(1/2, 0)$ and radius $\frac{1}{2}\cdot 5^{(k-1)/2}$ we have an ordered integral solution of the equation $(p - \frac{1}{2})^2 + (q - 0)^2 = 5^{k-1}/4$, and conversely.

We stress "ordered" here because the relation

$$\left(p - \frac{1}{2}\right)^2 + (q - 0)^2 = \frac{5^{k-1}}{4}$$

does not automatically imply that $(q - \frac{1}{2})^2 + (p - 0)^2 = 5^{k-1}/4$. Thus the number of ordered solutions (p, q) of this equation is the number of lattice points through which our circle passes. However, multiplying by 4, we see that each ordered solution (p, q) is also an ordered solution of the equation $(2p - 1)^2 + (2q)^2 = 5^{k-1}$, and conversely. But this equation is of the form $x^2 + y^2 = 5^{k-1}$. We are looking for solutions $x = 2p - 1$, $y = 2q$, that is, with x odd and y even. Each one we find yields a lattice point (p, q) on our circle. We have already observed that there are precisely $2k$ solutions (x, y) to $x^2 + y^2 = 5^{k-1}$ which have x odd and y even. Thus our circle must pass through precisely $2k = n$ lattice points.

When n is an odd number $2k + 1$, we consider solutions to

$$x^2 + y^2 = 5^{2k}.$$

The number of ordered integral solutions to this equation is

$$r(5^{2k}) = 4[(2k + 1) - 0] = 4(2k + 1) = 8k + 4.$$

These solutions generally go together in families of 8:

$$(x, y), (x, -y), (-x, y), (-x, -y),$$
$$(y, x), (y, -x), (-y, x), (-y, -x).$$

If a zero occurs, say $x = 0$, then $x = -x$, and the family reduces to just 4 solutions. Also, if $x = y$, there are only 4 solutions in the family.

In the case at hand we have 5^{2k} is an odd number, implying that one of x and y is odd and the other even. Thus we cannot have $x = y$. However, we can have a zero, giving the family

$$(0, 5^k), (0, -5^k), (5^k, 0), (-5^k, 0).$$

Consequently, our $8k + 4$ solutions (x, y) yield k families of 8 solutions and 1 family of 4 solutions.

Now the lattice points (p, q) lying on the circle with center

$(1/3, 0)$ and radius $\frac{1}{3}\cdot 5^k$ correspond 1-1 with the ordered integral solutions of the equations

$$\left(p - \frac{1}{3}\right)^2 + (q - 0)^2 = \frac{5^{2k}}{9} \quad \text{and} \quad (3p - 1)^2 + (3q)^2 = 5^{2k}.$$

This latter equation is of the form $x^2 + y^2 = 5^{2k}$. This time we seek solutions $x = 3p - 1$, $y = 3q$, that is, solutions with first term congruent to -1 mod 3 and second term divisible by 3. We shall find that each of the k families of 8 solutions (x, y) contains precisely two such solutions and that the single family of 4 solutions contains one solution, giving a total of $n = 2k + 1$ solutions, implying the same number of lattice points on our circle.

We note that

$$5^{2k} = 25^k \equiv 1^k \equiv 1 \pmod 3.$$

Since a square is congruent to 0 or 1 mod 3, it must be that one of x^2, y^2 is congruent to 1 and the other to 0 mod 3. Let x and $-x$ denote the terms in a family of 8 solutions which are divisible by 3. (We can arbitrarily label one solution of the family (x, y); the other labels are then determined $(-x, y)$, etc., in terms of this one.) In this case, either y or $-y$ is congruent to -1 mod 3. For definiteness, suppose it is y. Then only the two solutions (y, x) and $(y, -x)$ have first term congruent to -1 and second term congruent to 0 mod 3. In the family of 4 solutions, either $(-5^k, 0)$ or $(5^k, 0)$ is of the desired type. (QED)

3. Browkin's Theorem. The problems we have been considering take on an entirely different complexion when we try them with squares instead of circles. In 1957 George Browkin proved that, for every natural number n, there exists in the plane a square which contains in its interior exactly n lattice points. We consider Browkin's proof, as modified by Sierpinski and Schinzel.

Again we order the lattice points in a sequence $p_1, p_2, p_3, \ldots$. In order to do this we use the function

$$f(x, y) = |\, x + y\sqrt{3} - 1/3 \,| + |\, x\sqrt{3} - y - 1/\sqrt{3} \,|.$$

We shall show that no two lattice points give the same value of f. We proceed indirectly.

Suppose that for some distinct lattice points (a, b) and (c, d) we have $f(a, b) = f(c, d)$. Now $|z|$ is either $+z$ or $-z$, that is, $|z| = pz$ where p is either 1 or -1. Accordingly, removing the absolute values from the equation $f(a, b) = f(c, d)$, we obtain, where each of p, q, r, s is 1 or -1,

$$p(a + b\sqrt{3} - 1/3) + q(a\sqrt{3} - b - 1/\sqrt{3})$$
$$= r(c + d\sqrt{3} - 1/3) + s(c\sqrt{3} - d - 1/\sqrt{3}).$$

Separating rational and irrational terms, and noting that $s/\sqrt{3} = \sqrt{3}\cdot s/3$, we get

$$pa - p/3 - qb - rc + r/3 + sd$$
$$= \sqrt{3}(rd + sc - s/3 - pb - qa + q/3).$$

For equality, each side must vanish, giving

$$pa - qb - rc + sd + \frac{r - p}{3} = 0$$

$$\text{and} \quad rd + sc - pb - qa + \frac{q - s}{3} = 0.$$

Since all parameters denote integers, the fractions $(r - p)/3$ and $(q - s)/3$ must also reduce to integers. But each of the numerators is either 2, 0, or -2. Their only possible value, then, is 0. Thus $p = r$ and $q = s$, and the equations reduce to

$$p(a - c) + q(d - b) = 0 \quad \text{and} \quad p(d - b) + q(c - a) = 0.$$

Multiplying these equations, respectively, by p and q, we get

$$p^2(a - c) + pq(d - b) = 0 \quad \text{and} \quad pq(d - b) + q^2(c - a) = 0.$$

Subtraction now gives

$$p^2(a - c) - q^2(c - a) = 0, \text{ or } (a - c)(p^2 + q^2) = 0.$$

But whether p and q are 1 or -1, each of p^2 and q^2 is 1. Thus

$$2(a - c) = 0, \text{ giving } a = c.$$

This, in turn, yields $b = d$, identifying the lattice points (a, b) and (c, d); contradiction. Thus no two lattice points provide the same value of the function f.

Now f has a simple geometric interpretation. The length of the perpendicular d_1 from the lattice point $P(x, y)$ to the straight line L with equation $x + y\sqrt{3} - 1/3 = 0$ is given by

$$| d_1 | = \left| \frac{x + y\sqrt{3} - 1/3}{\sqrt{1+3}} \right|,$$

which yields $| x + y\sqrt{3} - 1/3 | = 2 | d_1 |$. Similarly,

$$| x\sqrt{3} - y - 1/\sqrt{3} | = 2 | d_2 |,$$

where d_2 denotes the perpendicular from $P(x, y)$ to the line M with equation $x\sqrt{3} - y - 1/\sqrt{3} = 0$.

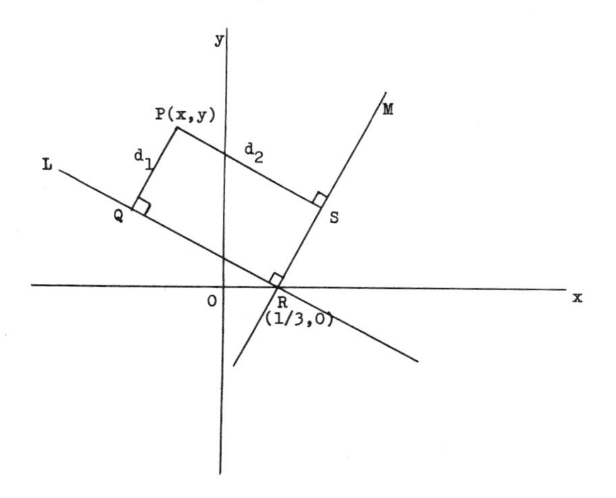

We note that the lines L and M meet at right angles at the point $R(1/3, 0)$. Thus $f(x, y) = 2 | d_1 | + 2 | d_2 |$ represents the perimeter of the rectangle $PQRS$ obtained by drawing d_1 and d_2 to L and M. Generally speaking, the value of f increases as $P(x, y)$ is taken farther away from the point of intersection R. Thus we see that some lattice point p_1 near R provides a rectangle of minimum perimeter and, since no two lattice points yield the same value of f, that increasing values of f order the lattice points in a sequence p_1, p_2, $p_3, \ldots$.

Let us use the notation a_n for the value of $f(x, y)$ determined by the lattice point p_n. Also let

$$h(x, y) = x(1 + \sqrt{3}) + y(\sqrt{3} - 1) - 1/3 - 1/\sqrt{3}, \quad \text{and}$$
$$g(x, y) = x(1 - \sqrt{3}) + y(1 + \sqrt{3}) - 1/3 + 1/\sqrt{3}.$$

Consider now the four straight lines

$$h(x, y) = \pm a_{n+1} \quad \text{and} \quad g(x, y) = \pm a_{n+1}.$$

Clearly the two lines related to $h(x, y)$ are parallel to each other, and also the other two are parallel to each other. Comparing the slopes of the lines from the equations, we see that the pairs of parallel lines are perpendicular. Thus they form a rectangle. In fact, they form a square. It is left to the reader to verify this by comparing the intercepts on the axes of the pairs of parallel lines.

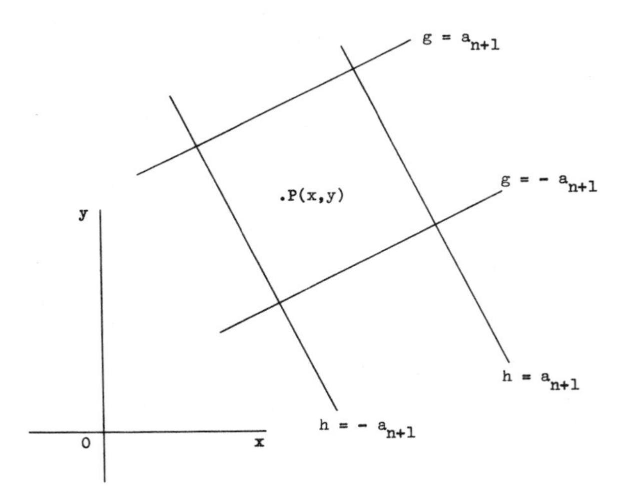

Now if a lattice point (x, y) provides a value of $h(x, y)$ such that

$$-a_{n+1} < h(x, y) < a_{n+1}, \quad \text{that is,} \quad |h(x, y)| < a_{n+1},$$

the point (x, y) will occur in the strip between the parallel lines

$h(x, y) = a_{n+1}$ and $h(x, y) = -a_{n+1}$, and conversely. Observing that the coefficient of x in $g(x, y)$ is negative, we see similarly that a lattice point occurs between the parallel lines $g(x, y) = a_{n+1}$ and $g(x, y) = -a_{n+1}$ if and only if we have $-a_{n+1} < g(x, y) < a_{n+1}$, that is, $|g(x, y)| < a_{n+1}$. The point (x, y), then, occurs in the square if and only if both $|h(x, y)| < a_{n+1}$ and $|g(x, y)| < a_{n+1}$.

Now, for real numbers a, b, c, the pair of inequalities $|a| < c$ and $|b| < c$ is equivalent to the single inequality

$$\left|\frac{a+b}{2}\right| + \left|\frac{a-b}{2}\right| < c$$

(see exercise 1). Accordingly, our two inequalities

$$|h(x, y)| < a_{n+1} \quad \text{and} \quad |g(x, y)| < a_{n+1}$$

are equivalent to

$$\left|\frac{h(x, y) + g(x, y)}{2}\right| + \left|\frac{h(x, y) - g(x, y)}{2}\right| < a_{n+1}.$$

Substituting the expressions for $h(x, y)$ and $g(x, y)$ and simplifying, we see easily that this reduces to just

$$f(x, y) < a_{n+1}.$$

That is to say, the lattice point (x, y) occurs in our square if and only if $f(x, y) < a_{n+1}$. Because of the order in our sequence, the lattice points satisfying this requirement are precisely

$$p_1, p_2, p_3, \ldots, p_n.$$

Hence our square contains exactly n lattice points. (QED)

Not only is the above theorem valid for squares. It holds also for triangles, pentagons, ellipses, and other figures. In fact, Schinzel and Kulikowski were able to prove the remarkable theorem that, for every nonempty plane bounded convex figure C, and for every natural number n, there exists in the plane a figure with the shape of C which contains in its interior exactly n lattice points!

4. Spheres in 3 Dimensions. We can generalize some of our results to spheres in 3 dimensions. For example, the proof that, for every natural number n, there is a sphere with exactly n lattice points in its interior is analogous to the proof of the corresponding plane theorem concerning circles. We conclude this chapter with an elegant proof of the following theorem.

KULIKOWSKI'S THEOREM. *For every natural number n, there exists in 3-dimensional space a sphere which has on its surface exactly n lattice points.*

This was proved by Thadée Kulikowski of Warsaw in 1958. The proof assumes a familiarity with Schinzel's theorem, which we proved earlier.

The number n is given. Using Schinzel's theorem, determine the equation of a circle,

$$(x-a)^2+(y-b)^2=c,$$

which passes through exactly n lattice points (x, y). Now a lattice point (x, y) in the xy-plane and the lattice point $(x, y, 0)$ in xyz-space are the same point. Consequently, this Schinzel circle contains exactly n lattice points $(x, y, 0)$ of 3-space.

Kulikowski now claims that the sphere with center $(a, b, \sqrt{2})$ and radius $\sqrt{c+2}$ has exactly n lattice points on its surface. Its equation is

$$(x-a)^2+(y-b)^2+(z-\sqrt{2})^2=c+2,$$

or

$$(x-a)^2+(y-b)^2+z^2-2z\sqrt{2}=c.$$

From Schinzel's theorem, the values of a, b, c are rational numbers ($a = 1/2$ or $1/3$, $b = 0$, and $c =$ the square of the appropriate radius). Consequently, in this equation, integral values of x, y, z give rational values for every term other than $-2z\sqrt{2}$. Thus there are no integers x, y, z satisfying the equation unless $z = 0$. That is to say, the lattice points (x, y, z) on the surface of the sphere all occur in its intersection with the xy-plane, obtained by putting $z = 0$. But this intersection is just the above Schinzel circle

$$(x-a)^2+(y-b)^2=c.$$

Thus the only lattice points on the sphere are the n lattice points $(x, y, 0)$ on this circle. The conclusion follows.

This proof generalizes immediately to a space of any number of dimensions. We note in closing that Browkin proved also that, for any natural number n, there is a cube in 3-space which contains exactly n lattice points in its interior.

Exercises

1. Prove that, for real numbers a, b, c, the inequalities $|a| < c$ and $|b| < c$ are equivalent to the single inequality

$$\left| \frac{a+b}{2} \right| + \left| \frac{a-b}{2} \right| < c.$$

2. Prove that, for every natural number n, there exists in 3-dimensional space a sphere containing exactly n lattice points in its interior. (Hint: Order the lattice points (x, y, z) by showing no two are the same distance from the point $(\sqrt{2}, \sqrt{3}, \sqrt{5})$.)

3. Prove, for every natural number n, that if there exists a circle in the plane with center (a, b), containing exactly n lattice points, then not both a and b can be rational numbers.

4. Prove that, for every natural number n, there exists in the plane a square which has exactly n lattice points on its boundary.

References and Further Reading

1. W. Sierpinski, A Selection of Problems in the Theory of Numbers, Pergamon Press, New York, 1964.

2. A. Schinzel, Sur L'Existence D'Un Cercle Passant Par Un Nombre Donné De Points Aux Coordonnées Entières, L'Enseignement Math., Series 2, vol. 4, (1958).

3. W. Sierpinski, Sur Quelques Problèmes Concernant Les Points Aux Coordonnées Entières, L'Enseignement Math., Series 2, vol. 4, (1958).

4. T. Kulikowski, Sur L'Existence D'Une Sphère Passant Par Un Nombre Donné Aux Coordonnées Entières, L'Enseignement Math., Series 2, vol. 5, (1959).

CHAPTER **12**

RECURSION

In this chapter we consider two problems whose solutions depend on recurrence relations. The first is very simple and almost seems to solve itself. Recursions and generating functions, which occur in our second problem, often surprise us with results that appear to be an inordinately high return on the bit of information we put in to start them off. They are very efficient techniques for extracting results from certain situations, and they are well worth knowing about.

1. The Gambler's Ruin. Our first problem is well known in the history of probability as the Gambler's Ruin. Suppose we repeatedly gamble pennies on the toss of a coin; heads you win, tails I win. Each of us has a probability of 1/2 of winning on each flip of the coin. Suppose, to begin with, you have 50 cents and I have 20 cents, and we play far into the night, if necessary, in order to complete our game. We ask for the three probabilities: (i) that you win all my money, (ii) that I win all your money, and (iii) that the game never ends.

Either of us can keep track of the game simply by counting his own money. Let us look at things from your point of view. It is convenient to mark the size of your bankroll on a number line extending through the integers 0, 1, 2, . . ., 70. To begin with you place a marker at 50. If you win the first toss, it is slid along to 51; if you lose, back to 49, and so on. Our first task is to find the probability of your marker reaching 70. That is to say, we would like to know the probability of beginning at 50 and winding up at

70. Of course, beginning at 49 and ending at 70 is essentially the same problem. Accordingly, let us define p_k to denote the probability of eventually reaching 70 from the starting point k. The probability of reaching 70 from the point k is the same whether one actually starts the game with k pennies or he acquires k pennies during the course of the game. The probability of reaching 70 from one's current position changes with each flip of the coin. When presently at k, p_k denotes the probability of finally winding up at 70, whether or not he oscillates through his present position during the course of the game. After one toss he is either at $k + 1$ or $k - 1$, and his new probability of reaching 70 is either p_{k+1} or p_{k-1}, respectively.

Now one way of winning the game from position k is to begin by moving to $k + 1$. The probability of this happening is 1/2. The probability of winning from there is p_{k+1}. Thus the probability of winning from k by first going to $k + 1$ is $\frac{1}{2}p_{k+1}$. Similarly, the probability of winning from position k by first moving to $k - 1$ is $\frac{1}{2}p_{k-1}$. These are the only two ways of winning from position k. Consequently, we have the recursion relation

$$p_k = \tfrac{1}{2}p_{k+1} + \tfrac{1}{2}p_{k-1},$$

and the problem is essentially solved.

Multiplying by 2 and transposing, we obtain

$$p_k - p_{k-1} = p_{k+1} - p_k.$$

This is a general formula, giving

$$p_1 - p_0 = p_2 - p_1 = p_3 - p_2 = \cdots = p_{70} - p_{69}.$$

Let us denote this constant difference by d. Adding these 70 differences, all but two terms cancel, giving

$$p_{70} - p_0 = 70d, \quad \text{and} \quad d = \tfrac{1}{70}(p_{70} - p_0).$$

Now the positions 70 and 0 represent the end of the game. At 0 you have nothing left to bet, so there is no chance of your ever reaching 70, implying $p_0 = 0$. At 70 I have nothing left to bet and you have already reached 70, implying $p_{70} = 1$. Thus

$$d = \tfrac{1}{70}(1 - 0) = \tfrac{1}{70}.$$

As a result, we have $p_0 = 0$, $p_1 = 1/70$, $p_2 = 2/70, \ldots, p_k = k/70$. Therefore your chance of winning all my money is 50/70. The analysis of the game is exactly the same from my point of view, giving the probability of my winning all your money to be 20/70.

We observe, then, that the probability of one of us cleaning out the other is

$$\frac{50}{70} + \frac{20}{70} = 1.$$

Thus it is a probabilistic certainty that one of us will ruin the other. The probability that the game never ends is 0. This is very surprising since there is an infinity of ways in which our fortunes could see-saw back and forth without either of us ever going broke.

2. A Problem of Euler. Drawing two nonintersecting diagonals in a convex pentagon divides the interior into three triangles. This can be done in five different ways. Three nonintersecting diagonals

can be drawn in a convex hexagon in 14 different ways to divide the interior into triangles. Our problem in this section is a difficult problem proposed by the great Swiss mathematician Leonhard Euler in the eighteenth century:

> In how many different ways can a convex n-gon be divided into triangles by nonintersecting diagonals?

A good notation is often extremely helpful in mathematics. It turns out that we are wise to use c_n to denote the required number of ways for an $(n + 2)$-gon. Thus $c_3 = 5$, in the case of pentagons, and $c_4 = 14$, for hexagons, and so on. Consider now a convex $(n + 3)$-gon, which gives rise to c_{n+1} decompositions. We denote the polygon $A_1A_2A_3 \cdots A_{n+2}A_{n+3}$.

In every decomposition, the side A_1A_{n+3} of our polygon occurs as a side in some triangle of the decomposition. The third vertex of such a triangle is one of the vertices $A_2, A_3, \ldots, A_{n+2}$. Let us work out the number of decompositions in which this third vertex is A_{k+2}. Then, letting k vary through the values $0, 1, \ldots, n$, we obtain c_{n+1} by adding all the numbers of decompositions thus obtained.

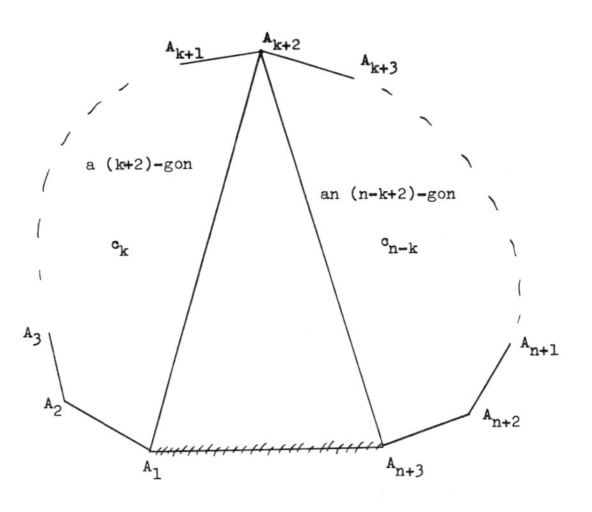

The diagonals from A_1 and A_{n+3} to the vertex A_{k+2} form a triangle which splits our $(n + 3)$-gon into two polygons, the $(k + 2)$-gon $A_1A_2\cdots A_{k+2}$ and the polygon $A_{k+2}A_{k+3}\cdots A_{n+3}$. This latter polygon has $n + 3 - (k + 1) = n - k + 2$ vertices. Accordingly, these polygons can, themselves, be decomposed, respectively, in c_k and c_{n-k} ways. A decomposition of our $(n + 3)$-gon is obtained for each of the c_k ways of decomposing the one polygon combined with each of the c_{n-k} ways of decomposing the other. With the vertices A_1 and A_{n+3} joined to A_{k+2}, then, there are altogether c_kc_{n-k} ways of decomposing the $(n + 3)$-gon. Summing over the values $k = 0, 1, 2, \ldots, n$, we find that the grand total c_{n+1} is given by

$$c_{n+1} = \sum_{k=0}^{n} c_k c_{n-k}.$$

However, we are getting a little ahead of ourselves here because the symbol c_0, as yet, is undefined. This symbol occurs only twice in the above sum. Let us look into these cases individually. For $k = 0$ and $k = n$, we obtain the terms $c_0 c_n$ and $c_n c_0$, respectively. These cases are depicted below.

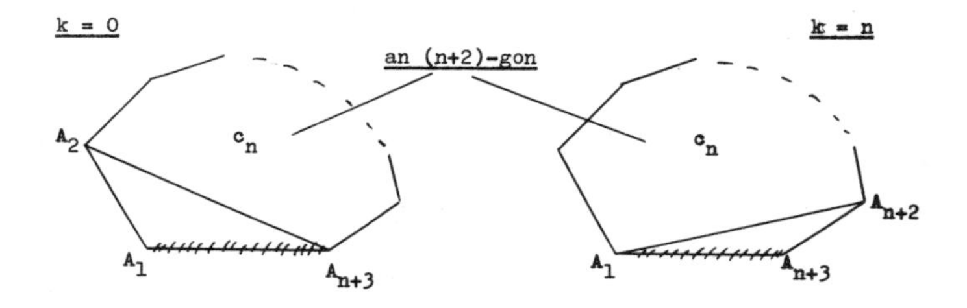

In each case, the one subpolygon collapses and the other is an $(n + 2)$-gon. The number of decompositions, then, is just c_n both times. This requires that we define $c_0 = 1$. We note in passing that $c_1 = 1$ also because there is only one way to decompose a triangle, namely by doing nothing, since it has no diagonals.

In order to solve our rather complicated recursion relation

$$c_{n+1} = \sum_{k=0}^{n} c_k c_{n-k},$$

we turn to generating functions. Let us define C to denote the infinite series in x whose coefficients are the unknown numbers c_0, c_1, etc., that is,

$$C = c_0 + c_1 x + c_2 x^2 + c_3 x^3 + \cdots + c_n x^n + \cdots,$$

or

$$C = \sum_{n=0}^{\infty} c_n x^n.$$

Now if C were to be multiplied by itself, the coefficient of x^n in the result would be

$$c_0c_n + c_1c_{n-1} + c_2c_{n-2} + \cdots + c_nc_0.$$

Thus we have that

$$C^2 = \sum_{n=0}^{\infty} [c_0c_n + c_1c_{n-1} + \cdots + c_nc_0]x^n.$$

By our recursion relation, this gives

$$C^2 = \sum_{n=0}^{\infty} c_{n+1}x^n.$$

Multiplying through by x, we get

$$xC^2 = \sum_{n=0}^{\infty} c_{n+1}x^{n+1}.$$

The right-hand side of this equation is just $c_1x + c_2x^2 + \cdots$, which is all of C except the first term c_0, which is 1. Thus

$$xC^2 = C - 1, \text{ giving } xC^2 - C + 1 = 0.$$

Solving this quadratic equation for C, we get

$$C = (1 \pm \sqrt{1 - 4x})/2x.$$

The binomial theorem gives

$$(1 - 4x)^{1/2} = 1 + \tfrac{1}{2}(-4x) + \cdots.$$

Consequently, the expression obtained by taking the $+$ sign, $(1 + \sqrt{1 - 4x})/2x$, has first term $1/x$. But C has no term in $1/x$. Thus this expression must be extraneous, and the series for C is given by

$$C = \frac{1 - \sqrt{1 - 4x}}{2x} = \frac{1}{2x}[1 - \sqrt{1 - 4x}].$$

This gives

$$C = \frac{1}{2x}\left[-\frac{1}{2}(-4x) - \frac{\frac{1}{2}(\frac{1}{2} - 1)}{1.2}(-4x)^2 + \cdots\right],$$

where every term in the bracket is preceded by a minus sign. In the series for C, the desired number c_n is the coefficient of x^n. We have

$$c_n x^n = \frac{1}{2x}\left[-\frac{\frac{1}{2}(\frac{1}{2}-1)\cdots[\frac{1}{2}-(n+1)+1]}{1\cdot 2\cdot 3\cdots n(n+1)}(-4x)^{n+1}\right].$$

Thus

$$c_n = \frac{1}{2}\left[-(-1)^n\frac{\frac{1}{2}(1-\frac{1}{2})\cdots[(n+1)-1-\frac{1}{2}]}{1\cdot 2\cdot 3\cdots n(n+1)}(-1)^{n+1}4^{n+1}\right]$$

$$= \frac{1}{2}\left[(-1)^{2n+2}\frac{\frac{1}{2}\cdot\frac{1}{2}\cdot\frac{3}{2}\cdot\frac{5}{2}\cdots(2n-1)/2}{1\cdot 2\cdot 3\cdots n(n+1)}2^{2n+2}\right]$$

$$= \frac{1\cdot 1\cdot 3\cdot 5\cdots(2n-1)}{1\cdot 2\cdot 3\cdots n(n+1)}2^n.$$

Multiplying top and bottom of this fraction by $n!$, we obtain the neat expression

$$c_n = \frac{1\cdot 3\cdot 5\cdots(2n-1)}{(n+1)!\,n!}[(2\cdot 1)(2\cdot 2)(2\cdot 3)\cdots(2\cdot n)]$$

$$= \frac{1\cdot 3\cdot 5\cdots(2n-1)[2\cdot 4\cdot 6\cdot 8\cdots 2n]}{(n+1)!\,n!},$$

i.e.,

$$c_n = \frac{(2n)!}{n!\,(n+1)!}.$$

Exercises

1. Generating functions are useful in solving many recurrence relations. We consider an example in detail.

If f_n denotes the number of parts into which the plane is divided by n straight lines (no two of which are parallel and no three concurrent), we obtain

$$f_{n+1} = f_n + (n + 1), \quad \text{where } f_0 = 1.$$

Defining $F(x) = f_0 + f_1x + f_2x^2 + \cdots + f_nx^n + \cdots$, we get $x \cdot F(x) = f_0x + f_1x^2 + \cdots + f_{n-1}x^n + \cdots$. Subtraction gives

$$\begin{aligned}
(1 - x)F(x)& \\
&= f_0 + (f_1 - f_0)x + (f_2 - f_1)x^2 + \cdots + (f_n - f_{n-1})x^n + \cdots \\
&= 1 + 1 \cdot x + 2x^2 + 3x^3 + \cdots + nx^n + \cdots \\
&= 1 + x(1 + 2x + 3x^2 + \cdots + nx^{n-1} + \cdots) \\
&= 1 + x(1 - x)^{-2}.
\end{aligned}$$

Thus $F(x) = (1 - x)^{-1} + x(1 - x)^{-3}$.

On the left-hand side, the coefficient of x^n is f_n, by definition. This is determined, then, by finding the coefficient of x^n on the right-hand side. Thus

$$f_n = (\text{coeff. of } x^n \text{ in } (1 - x)^{-1}) + (\text{coeff. of } x^n \text{ in } x(1 - x)^{-3}).$$

(Observe that the latter term here is just the coefficient of x^{n-1} in $(1 - x)^{-3}$.) The binomial theorem yields the result.

In doing this work it is very useful to recognize the frequently occurring series for

$$(1 - x)^{-1}, (1 + x)^{-1}, (1 - x)^{-2}, (1 + x)^{-2}, (1 - x)^{-3}, (1 + x)^{-3}.$$

In the problem at hand, we get $f_n = 1 + n(n + 1)/2$, giving

$$f_n = \frac{n^2 + n + 2}{2}.$$

Apply this approach to the recurrence $f_{n+1} = f_n + 2n$, for $n = 1, 2, 3, \ldots$, and $f_0 = 1, f_1 = 2$.

2. In decomposing a convex n-gon into triangles by drawing non-intersecting diagonals, how many triangles are formed? How many diagonals are drawn?

3. A fly walks along the edges of the cube $ABCDEFGH$, starting from the vertex A. The probability that the fly chooses to go along

a particular edge from a vertex is 1/3. The vertices F and G are covered with fly-paper. What is the probability that the fly gets stuck at vertex G? What is the probability the fly gets stuck at F? What is the probability that he doesn't get stuck at all?

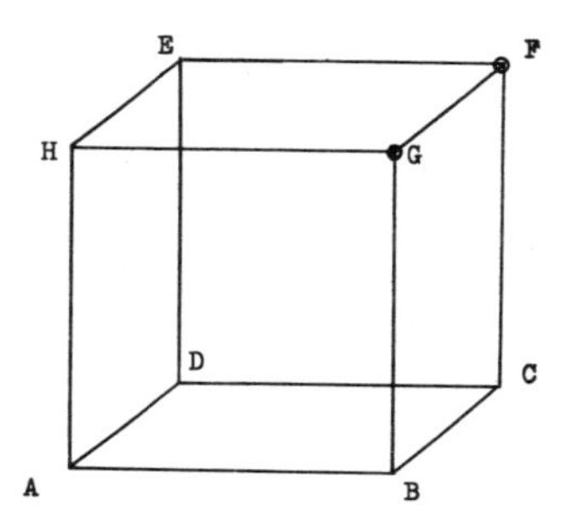

4. Use generating functions to determine a formula for the nth term of the famous Fibonacci sequence, which is defined by

$$f_{n+2} = f_{n+1} + f_n, \qquad f_1 = f_2 = 1, \qquad n \geqq 1.$$

(Hint: Define $F(x) = \sum_{n=0}^{\infty} f_{n+1}x^n$. Determine $xF(x)$ and $x^2F(x)$ and obtain the value of $(1 - x - x^2)F(x)$.) (The formula thus obtained is called Binet's formula.)

5. A running total of the values obtained by throwing a die is kept. What is the probability that a specified integer n occurs in the sequence thus obtained?

6. $2n$ points are taken on a circle. How many ways can they be joined in pairs by n nonintersecting chords?

References and Further Reading

1. Dynkin and Uspenskii, Random Walks, Heath, Boston, 1963.

2. Yearbook 28 of the NCTM, Enrichment Mathematics for High Schools, 1963.

CHAPTER **13**

POULET, SUPER-POULET, AND RELATED NUMBERS

1. Poulet and Super-Poulet Numbers. In 1926 P. Poulet published a table of odd pseudoprimes up to 50 million. In 1938 he extended it to 100 million. Accordingly, pseudoprimes have come to be known as Poulet numbers. We recall that a pseudoprime[1] is a composite integer n such that $n \mid 2^n - 2$. We see that 2047 is a Poulet number as follows.

First of all $2047 = 2^{11} - 1$. Also $2047 = 11 \cdot 186 + 1$. Thus

$$2^{2047} - 2 = 2^{11 \cdot 186 + 1} - 2 = 2(2^{11 \cdot 186} - 1) = 2[(2^{11})^{186} - 1^{186}]$$

$$= 2(2^{11} - 1)(\cdots) = 2(2047)(\cdots).$$

This Poulet number has the property that all its divisors d also satisfy the defining relation $d \mid 2^d - 2$. The prime decomposition of 2047 is $23 \cdot 89$, implying that the divisors of 2047 are just these two primes, themselves. Fermat's Simple Theorem assures us that they satisfy $d \mid 2^d - 2$. A Poulet number, all of whose divisors d satisfy $d \mid 2^d - 2$, is called a **super-Poulet** number. As we have seen, Fermat's theorem ensures the relation for all prime divisors. Thus we may alternatively define a super-Poulet number as a Poulet number whose composite divisors are also Poulet numbers.

Not all Poulet numbers turn out to be super-Poulet numbers. For example, for the Poulet number 561 we have $561 = 3 \cdot 11 \cdot 17$, with divisor 33; however $33 \nmid 2^{33} - 2$. To see this, we note that

[1] See the essay "An Old Chinese Theorem and Pierre de Fermat".

$2^{10} = 1024 = 11 \cdot 93 + 1 \equiv 1 \pmod{11}$. Therefore $2^{30} \equiv 1 \pmod{11}$. However, $2^3 \equiv 8 \pmod{11}$, leading to $2^{33} \equiv 8 \pmod{11}$ and $2^{33} - 2 \equiv 6 \pmod{11}$. Thus $11 \nmid 2^{33} - 2$, implying that $33 \nmid 2^{33} - 2$. Thus some Poulet numbers are super-Poulet and others are not. It turns out that there is an infinity of each.

In 1936 the American mathematician D. H. Lehmer proved that an infinity of Poulet numbers have just two prime factors, as 2047, thus assuring an infinity of super-Poulet numbers. On the other hand, no even number can be super-Poulet, and Beeger's theorem (1951) established that there is an infinity of even Poulet numbers. Let us show that all super-Poulet numbers are odd.

Suppose, to the contrary, that an even number $2n$ is a super-Poulet number. In this case, we have

$$\text{(i)}\ 2n \mid 2^{2n} - 2, \quad \text{and, for the divisor } n, \quad \text{(ii)}\ n \mid 2^n - 2.$$

From (i) we see, dividing through by 2, that $n \mid 2^{2n-1} - 1$. Thus n must be odd. Because of this, relation (ii), which is

$$n \mid 2(2^{n-1} - 1), \quad \text{gives} \quad n \mid 2^{n-1} - 1.$$

Dividing each term, we see that n divides the difference $(2^{2n-1} - 1) - (2^{n-1} - 1)$, that is,

$$n \mid 2^{2n-1} - 2^{n-1}, \text{ or } n \mid 2^{n-1}(2^n - 1).$$

Since n is odd, this means that $n \mid 2^n - 1$. Since n also divides $2^n - 2$, it must divide their difference, namely 1. Thus $n = 1$, and $2n = 2$. In this case, the prime 2 is a super-Poulet number, and therefore a Poulet number. This is impossible since Poulet numbers are composite, by definition. (QED)

2. Related Numbers. Fermat's theorem states that a prime number n divides $a^n - a$ for all integers a. In the case of an integer a which is relatively prime to n, we have

$$n \mid a^n - a = a(a^{n-1} - 1), \quad \text{giving} \quad n \mid a^{n-1} - 1.$$

Composite numbers n for which $n \mid a^{n-1} - 1$ whenever a and n are relatively prime (i.e., $(a, n) = 1$) were first noticed by Robert Carmichael in 1909, and are called Carmichael numbers. Clearly the absolute pseudoprimes (i.e., composite n such that $n \mid a^n - a$ for all integers a) are Carmichael numbers. The converse is also true, showing the Carmichael numbers and the absolute pseudoprimes to be identical.

There are composite numbers n such that $n \mid a^{n-2} - a$ whenever $(a, n) = 1$. Such a number is $n = 195$. The prime decomposition of 195 is $3 \cdot 5 \cdot 13$. Each of these primes divides $a^{193} - a$ for all integers a. We consider the case of the prime 5; the others are similar.

$$193 = (5 - 1) \cdot 48 + 1 = 4 \cdot 48 + 1;$$

thus

$$a^{193} - a = a^{4 \cdot 48+1} - a = a[(a^4)^{48} - 1^{48}] = a(a^4 - 1)(\cdots)$$
$$= (a^5 - a)(\cdots).$$

Since 5 is a prime, Fermat's theorem gives $5 \mid a^5 - a$, assuring that $5 \mid a^{193} - a$.

In the case of relatively prime a and n, the relation

$$n \mid a^{n-2} - a = a(a^{n-3} - 1)$$

gives $n \mid a^{n-3} - 1$. Natural numbers n greater than 3 (not just composite ones) such that

$$n \mid a^{n-3} - 1 \qquad \text{whenever} \quad (a, n) = 1$$

are called D-numbers. They were studied in 1951 by D. C. Morrow. We prove that there exists an infinity of D-numbers by showing that three times an odd prime is always a D-number.

The odd prime $p = 3$ is considered separately. In this case $n = 3p = 9$ and we show that $9 \mid a^6 - 1$ for all integers a such that $(a, 9) = 1$. Because a and 9 are relatively prime we have

$$a \equiv \pm 1, \pm 2, \pm 4 \pmod 9.$$

In each case we easily obtain $a^6 \equiv 1 \pmod 9$, as desired.

Suppose, then, that $n = 3p$ where p is an odd prime greater than 3. We show that

$$n = 3p \mid a^{3p-3} - 1 \qquad \text{whenever} \quad (a, n) = (a, 3p) = 1.$$

Because $(a, 3p) = 1$, a is not a multiple of 3. Then $a \equiv \pm 1 \pmod 3$, giving $a^2 \equiv 1 \pmod 3$, and $a^{2k} \equiv 1 \pmod 3$ for all natural numbers k. That is to say, every even power of a is congruent to 1 mod 3. Since p is odd, we have $3p - 3$ even, implying that $a^{3p-3} - 1$ is divisible by 3.

Since p exceeds 3 and p is a prime we have $(3, p) = 1$. Thus if each of 3 and p divides $a^{3p-3} - 1$, their product n does also. We conclude the proof by showing that p divides $a^{3p-3} - 1$. We have

$$a^{3p-3} - 1 = (a^{p-1})^3 - 1^3 = (a^{p-1} - 1)(\cdots).$$

By Fermat's Simple Theorem, we see for $(a, p) = 1$ that $p \mid a^{p-1} - 1$. The conclusion follows.

In 1962, A. Makowski showed that there exists for all natural numbers $k \geqq 2$ an infinity of composite numbers n such that

$$n \mid a^{n-k} - 1 \qquad \text{whenever} \quad (a, n) = 1.$$

For $k = 3$, this theorem confirms an infinity of composite D-numbers. For $k = 2$, this gives an infinity of composite n such that

$$n \mid a^{n-2} - 1 \qquad \text{whenever} \quad (a, n) = 1.$$

For such n, we see that $n \mid a^{n-1} - a$ whenever $(a, n) = 1$. However, we can say even more: there is an infinity of composite numbers n such that $n \mid a^{n-1} - a$ for all integers a, whether or not a and n are relatively prime. We prove that if p is an odd prime, then $n = 2p$ is such a number.

Clearly a and a^{n-1} are both odd or both even, implying that $2 \mid a^{n-1} - a$. Since p is an odd prime, we have $(2, p) = 1$, implying that n divides $a^{n-1} - a$ if both 2 and p do. Now

$$\begin{aligned} a^{n-1} - a &= a^{2p-1} - a \\ &= a(a^{2p-2} - 1) = a[(a^{p-1})^2 - 1^2] \\ &= a(a^{p-1} - 1)(a^{p-1} + 1) = (a^p - a)(a^{p-1} + 1). \end{aligned}$$

Fermat's theorem gives $p \mid a^p - a$, completing the proof.

3. Mersenne and Related Numbers. The numbers $M_n = 2^n - 1$ are called Mersenne numbers after a French cleric, Marin

Mersenne (1588–1648), whose voluminous correspondence with leading scholars did much to make known important mathematical results at a time when technical journals were nonexistent. As noted in the chapter on perfect numbers, they occur in Euclid's formula for even perfect numbers:

$$m = 2^{n-1}(2^n - 1) \qquad \text{is perfect if } 2^n - 1 \text{ is prime.}$$

Since the converse is also true, proved by Euler in the 18th century, the quest for even perfect numbers is identical to the search for Mersenne numbers which are prime, the so-called Mersenne primes.

It is an easy exercise to show that n is prime if M_n is prime. One is tempted to assume the converse. However, as we have seen, $2^{11} - 1$ is the pseudoprime 2047. Mersenne, himself, claimed that M_n is prime for $n = 2, 3, 5, 7, 13, 17, 31, 67, 127, 257$ and that it is composite for all other primes $n \leqq 257$. Later developments have shown his statement to be incorrect. He should have included $n = 19, 61, 89$, and 107, and he should have left out $n = 67$ and 257.

In view of the interest in Mersenne numbers generated by their relation to perfect numbers, it is no surprise that, since the time people began to keep track of such things, the world's record for the greatest known prime number has always been a Mersenne prime. The prime $M_{127} = 2^{127} - 1$, a number of 39 digits, held the title from 1914 to the early fifties when, in just a few months, it was passed through M_{521}, M_{607}, M_{1279}, M_{2203} to M_{2281}, a number of 687 digits. Since this initial surge of results due to electronic computers, a great many factorizations have been accomplished and a further seven Mersenne primes have been discovered—M_n for $n = 3217, 4253, 4423, 9689, 9941, 11213$—culminating with M_{19937}, found to be prime by Bryant Tuckerman on the evening of March 4, 1971. This brings the number of known perfect numbers to 24. It is not known whether there exists an infinity of Mersenne primes.

Now, by definition, a Poulet number n divides $2^n - 2$. However, for the closely related Mersenne numbers, there exist no integers $n > 1$ such that $n \mid 2^n - 1$. The following proof is due to the outstanding Polish mathematician A. Schinzel.

He proceeds indirectly. Suppose that $n > 1$ does divide $2^n - 1$. Since $2^n - 1$ is odd, then its divisor n must be odd, too. Let p denote the smallest prime divisor of n. Then p also is odd. Accordingly,

$$(p, 2) = 1,$$

and we have $p \mid 2^{p-1} - 1$ by Fermat's theorem.

We look now at the values of m for which $p \mid 2^m - 1$. We see that $m = p - 1$ is one value. It may happen that there are also values of m which are less than $p - 1$. Let q denote the smallest value of m. Then we know that $q \leqq p - 1$ and $p \mid 2^q - 1$. Since p is prime, it is greater than 1. In order for $2^q - 1$ to be divisible by p then, it must be that q exceeds 1. Thus we have

$$1 < q \leqq p - 1, \quad \text{which gives} \quad 1 < q < p.$$

We obtain a contradiction by showing that q divides n, implying that p is not the smallest prime divisor of n (any prime divisors of q are smaller).

We even prove this indirectly. Suppose that q does not divide n. Then we would have

$$n = kq + r \quad \text{for some integer } k, \text{ where} \quad 0 < r < q.$$

Now $p \mid 2^q - 1$, or $2^q \equiv 1 \pmod{p}$. Then

$$2^n - 1 = 2^{kq+r} - 1 = 2^r(2^q)^k - 1 \equiv 2^r(1)^k - 1 \equiv 2^r - 1 \pmod{p}.$$

Because $n \mid 2^n - 1$ and $p \mid n$, we have $p \mid 2^n - 1$, or $2^n - 1 \equiv 0 \pmod{p}$. Therefore we have

$$2^r - 1 \equiv 0 \pmod{p}, \quad \text{or} \quad p \mid 2^r - 1.$$

But $r < q$, contradicting the minimality of q as a value of m. (QED)

It is not difficult to show that there is an infinity of natural numbers n such that $n \mid 2^n + 1$. In fact, $n = 3^k$, for $k = 0, 1, 2, \ldots$, is such a number. The proof, which is left to the reader, is an easy application of induction.

Finally, while there exists an infinity of natural numbers n such that $n \mid 2^n + 2$ (e.g., $n = 2$, 6, and 66), there are no natural $n > 1$ such that $n \mid 2^{n-1} + 1$.

Exercises

1. Prove that $n^2 + 3n + 5$ is never divisible by 121 for any natural number n.
2. Show that, if $2^n - 1$ is a prime number, then n is prime. (Note that the converse fails for $n = 11$.)
3. Prove that $2^n - 1$ and $2^m + 1$ are always relatively prime, m and n natural numbers, n odd.
4. What is the least positive integer n such that n divides $3^n - 3$ but does not divide $2^n - 2$? What is the least n which divides $2^n - 2$ but not $3^n - 3$?
5. Find all prime numbers p such that p divides $2^p + 1$.
6. Given the positive integer a, find a composite natural number n such that n divides $a^n - a$.

References and Further Reading

1. W. Sierpinski, 250 Problems in Elementary Number Theory, American Elsevier, New York, 1970.
2. ———, A Selection of Problems in The Theory of Numbers, Pergamon Press, New York, 1963.
3. ———, Elementary theory of numbers, Warszawa (1964).
4. D. H. Lehmer, On the converse of Fermat's theorem, Amer. Math. Monthly, 43 (1936) 347–356.

SOLUTIONS TO SELECTED EXERCISES

1. An Old Chinese Theorem and Pierre De Fermat.

6. Clearly $2^n - 1$ is odd and greater than n. Since n is odd and composite, $n = ab$ for some pair of odd natural numbers a and b, each of which exceeds 1. Then

$$2^n - 1 = 2^{ab} - 1 = (2^a)^b - 1^b = (2^a - 1)(\cdots).$$

Here $2^a - 1$ is not 1, lest $a = 1$, and it is not $2^n - 1$, lest $b = 1$. Thus $2^n - 1$ is composite.

Since n is a pseudoprime, n divides $2^n - 2$. Thus $2^n - 2 = nk$, for some natural number k. Hence

$$2^{(2^n-2)} - 1 = 2^{nk} - 1 = (2^n)^k - 1^k = (2^n - 1)(\cdots).$$

Thus $2^n - 1$ divides $2^{(2^n-2)} - 1$. Accordingly, it also divides twice as much, namely $2^{(2^n-1)} - 2$, completing the proof that $2^n - 1$ is a pseudoprime.

7. Let $(2^m - 1, 2^n + 1) = d$. For some natural numbers k and q we have $2^m - 1 = kd$ and $2^n + 1 = qd$. These yield $2^m = kd + 1$ and $2^n = qd - 1$. Thus $2^{mn} = (kd + 1)^n = (qd - 1)^m$. Expanding by the binomial theorem, we see that, for some integers t and s, we have

$$(kd + 1)^n = td + 1 \quad \text{and} \quad (qd - 1)^m = sd - 1,$$

(recall that m is odd). Thus $td + 1 = sd - 1$, giving $2 = d(t - s)$, implying that d divides 2. Thus $d = 1$ or $d = 2$. But $2^m - 1$ is odd, implying d is odd. Thus $d = 1$.

2. Louis Pósa.

6. In a $4 \times n$ chessboard there are two rows of outer squares $O_1, O_2, \ldots, O_{2n}$, and two rows of inner squares $I_1, I_2, \ldots, I_{2n}$. A re-entrant knight's tour puts these $4n$ squares in a cycle.

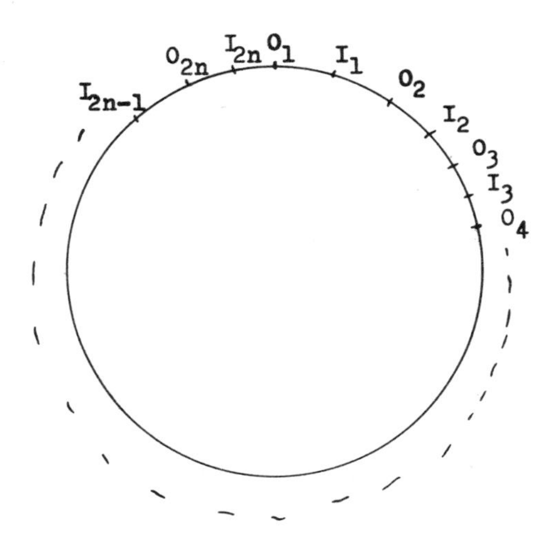

Now a knight on an outer square, by the nature of its move, must move to an inner square. Thus, in the cycle, each of the outer squares must be followed by an inner square. This forces the outer and inner squares to alternate in the cycle.

However, each time a knight moves, it changes color. Thus the colors of the squares in the cycle alternate. This makes all the outer squares one color and all the inner squares the other color; impossible.

10. The solution turns on the fact that the edges A_pA_q, A_rA_s are parallel if and only if $p + q \equiv r + s \pmod{2n}$. This follows easily from the observation that parallel edges cut off equal intercepts on the regular $2n$-gon, and conversely (the edges are parallel if and only if the number of vertices x between A_q and A_r at the one side is the same as the number between A_p and A_s at the other side).

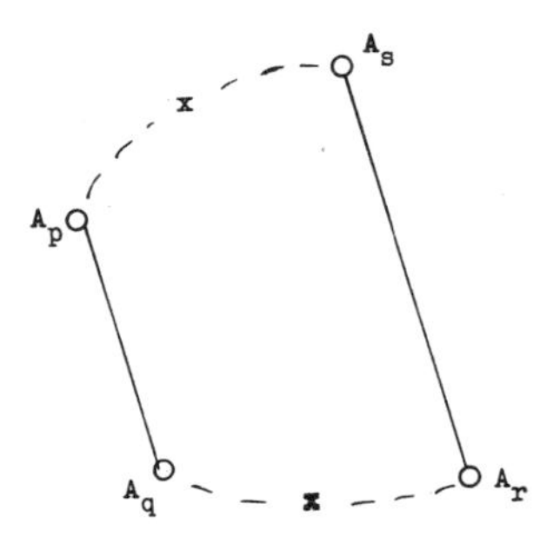

On this basis, the problem is solved indirectly as follows.

Suppose $A_{a_0}A_{a_1}A_{a_2}\cdots A_{a_{2n-1}}$ denotes a Hamiltonian circuit in which no two edges are parallel. For no two edges, then, can the sums of the subscripts be congruent mod $2n$. Thus no two of the $2n$ numbers $a_0 + a_1$, $a_1 + a_2$, . . ., $a_{2n-1} + a_0$ can be congruent mod $2n$. Among these numbers, then, each of the $2n$ residues 0, 1, 2, . . ., $2n - 1$ must occur once and only once.

$$\text{Thus } (a_0 + a_1) + (a_1 + a_2) + \cdots + (a_{2n-1} + a_0)$$

$$\equiv 0 + 1 + \cdots + 2n - 1 = \frac{(2n-1)(2n)}{2} = 2n^2 - n$$

$$\equiv -n \equiv n \pmod{2n}.$$

$$\text{However, } (a_0 + a_1) + (a_1 + a_2) + \cdots + (a_{2n-1} + a_0)$$

$$= 2(a_0 + a_1 + a_2 + \cdots + a_{2n-1})$$

$$= 2(0 + 1 + 2 + \cdots + 2n - 1) \text{ (in some order)}$$

$$= (2n-1)(2n)$$

$$\equiv 0 \pmod{2n}.$$

This contradiction concludes the proof.

3. Equilateral Triangles.

5. Let circles DAB, FAC meet at X. Then $\angle AXB = 180° - D$, $\angle AXC = 180° - F$. Thus $\angle BXC = 360° - \angle AXB - \angle AXC = D + F = 180° - E$. Thus angles BXC and E are supplementary, implying $BXCE$ is cyclic. Thus the three circles are concurrent. Also, the line of centers of two intersecting circles is

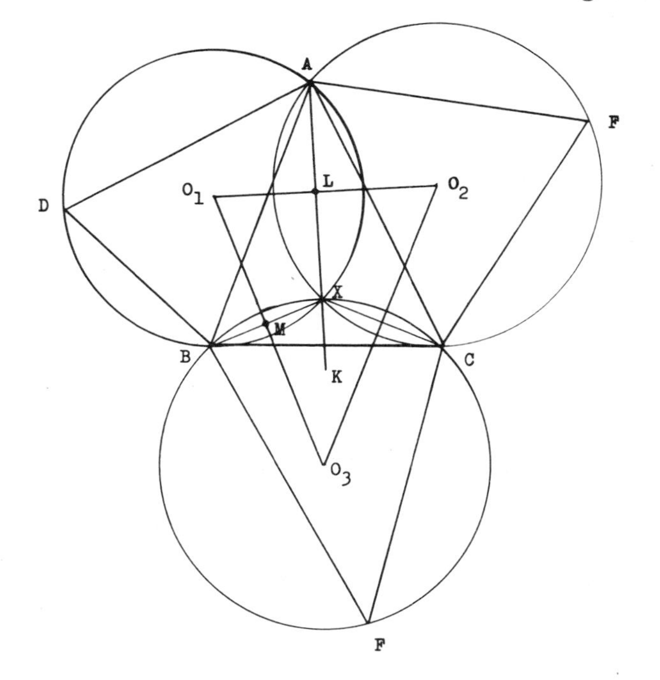

perpendicular to the common chord. Thus O_1O_2 and O_1O_3 are respectively perpendicular to AX and BX. Thus $\angle O_2O_1O_3$ equals the clockwise angle between AX and BX, beginning with AX. This makes it equal to the exterior angle at X of cyclic quadrilateral $ADBX$. Hence $\angle O_2O_1O_3 = D$. Similarly for E and F. Thus $\Delta O_1O_2O_3$ has the angles D, E, F. Consequently, if $D = E = F = 60°$, $\Delta O_1O_2O_3$ is equilateral. This is Napoleon's Theorem.

6.

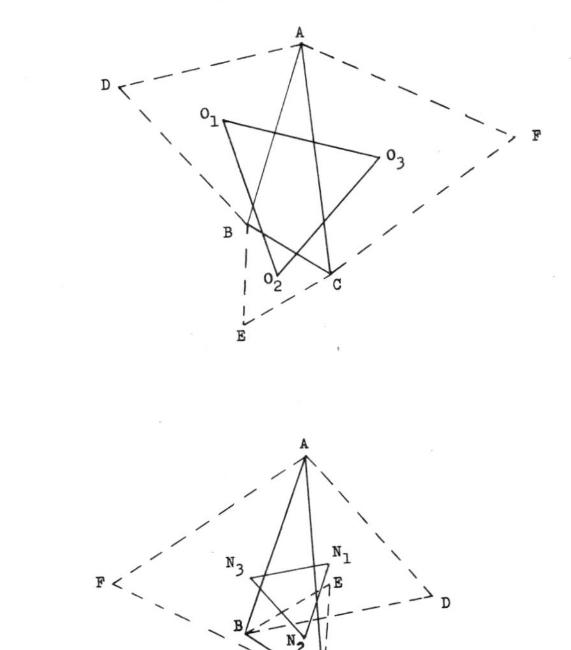

The outer and inner cases are drawn (supposed to be for the same ΔABC). Let the triangles of centers be $O_1O_2O_3$ and $N_1N_2N_3$. We have established already that the outer triangle is equilateral. In the outer case, we have, applying the law of cosines to ΔAO_1O_3

$$O_1O_3^2 = AO_1^2 + AO_3^2 - 2\cdot AO_1\cdot AO_3\cdot\cos\angle O_1AO_3.$$

Let the sides of ΔABC be a, b, c, as usual. Then AO_1 is the radius of the circumcircle around ΔABD, which is an equilateral triangle with side c. Accordingly, the length of AO_1 is two-thirds an altitude of this triangle, namely $\frac{2}{3}(c\sqrt{3}/2) = c/\sqrt{3}$. Thus $AO_1^2 = c^2/3$. Similarly $AO_3^2 = b^2/3$.

Now the angles O_1AB and O_3AC are 30 degrees, making $\angle O_1AO_3 = A + 60°$. Hence

$$O_1O_3^2 = \frac{c^2}{3} + \frac{b^2}{3} - \frac{2}{3}\,bc\cdot\cos(A + 60°).$$

In the inner case, doing the same thing to ΔAN_1N_3 we get:

$$N_1N_3^2 = \frac{c^2}{3} + \frac{b^2}{3} - \frac{2}{3}\, bc\cdot\cos\,\angle N_1AN_3.$$

Now the inner diagram is obtained from the outer diagram by reflecting the equilateral triangles in their respective sides of the ΔABC. In this way O_1 goes into N_1, etc. As our diagrams are labeled, AO_1, in this reflection, swings through an angle of 60° about the point A, in reaching its image AN_1. This diminishes the angle O_1AO_3 by 60°. Similarly, AO_3 in swinging to AN_3 diminishes the angle O_1AO_3 by 60°. Thus the angle N_1AN_3 is 120° less than $\angle O_1AO_3$, making it $A - 60°$. If A is less than 60° to begin with (as in our figures), then the angle N_1AN_3 becomes negative. But this is all right. Subtracting the above results, we get

$$O_1O_3^2 - N_1N_3^2 = \tfrac{2}{3}bc[\cos(A - 60°) - \cos(A + 60°)]$$

$$= \frac{2}{3}\, bc\cdot 2\cdot\sin A\cdot\sin 60° = \frac{2}{\sqrt{3}}\, bc\cdot\sin A.$$

(Observe that the cosine of a negative angle is the same as the cosine of the positive angle with the same magnitude.) Thus

$$O_1O_3^2 - N_1N_3^2 = \frac{4}{\sqrt{3}}\left[\frac{1}{2}\, bc\cdot\sin A\right] = \frac{4}{\sqrt{3}}\,(\Delta ABC).$$

This is independent of the sides chosen. Thus $O_2O_3^2 - N_2N_3^2$ and $O_1O_2^2 - N_1N_2^2$ are the same amount. Since these are equal to each other, and the O_1O_2, O_2O_3, O_3O_1 are equal, we see that the N_1N_2, N_2N_3, N_3N_1 are equal. (QED)

7. This follows as a corollary to 6. The area of an equilateral triangle with side x is $\sqrt{3}\cdot x^2/4$. Thus $\Delta O_1O_2O_3 = (\sqrt{3}/4)\cdot O_1O_3^2$, and $\Delta N_1N_2N_3 = (\sqrt{3}/4)\cdot N_1N_3^2$. From 6, we have

$$\frac{\sqrt{3}}{4}\cdot O_1O_3^2 - \frac{\sqrt{3}}{4}\cdot N_1N_3^2 = \Delta ABC,$$

which is the required result.

8. The triangle AO_1N_3 is similar to ΔABC and the ratio of corresponding sides is $1/\sqrt{3}$. We have established before that $AO_1 = c/\sqrt{3}$ and $AO_3 = b/\sqrt{3}$. (This makes $AN_3 = b/\sqrt{3}$.) And the angle $O_3AO_1 = A + 60°$. However, the angle N_3AO_1 is just $60°$ smaller (lost in reflecting AO_3 in AB to get AN_3). Hence $\angle N_3AO_1 = A$. (QED)

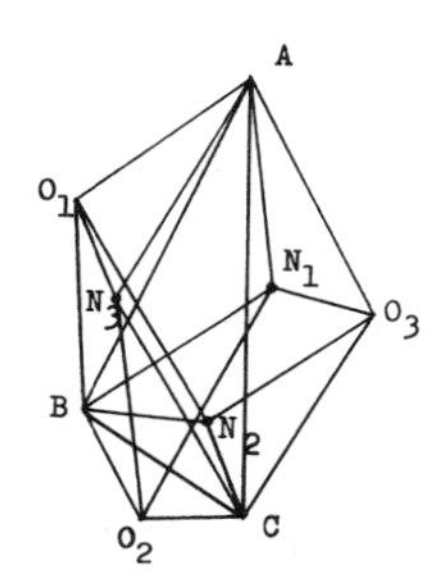

Similarly, there are five other triangles similar to ΔABC with corresponding sides in the same ratio $1/\sqrt{3}$. Thus they are all congruent to each other. They are AN_1O_3, BN_1O_2, BN_2O_1, CN_2O_3, CN_3O_2. The reflections producing N_1, N_2, N_3 yield six equilateral triangles: AO_1N_1, BO_1N_1, AO_3N_3, CO_3N_3, BO_2N_2, CO_2N_2. Their equal sides are in each case $1/\sqrt{3}$ times a side of ΔABC. Thus from the congruent and equilateral triangles we have

$$AO_1 = AN_1 = BO_1 = BN_1 = N_3O_2 = N_2O_3 = \frac{c}{\sqrt{3}},$$

$$AO_3 = AN_3 = CO_3 = CN_3 = N_2O_1 = N_1O_2 = \frac{b}{\sqrt{3}},$$

$$O_1N_3 = O_3N_1 = BO_2 = BN_2 = CO_2 = CN_2 = \frac{a}{\sqrt{3}}.$$

Now quadrilateral $O_1BO_2N_3$ has opposite sides equal. This makes it a parallelogram.

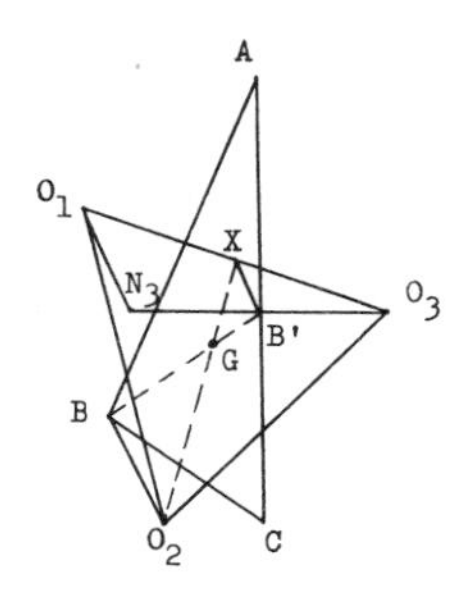

Let O_2X be a median of $\Delta O_1O_2O_3$, and let BB' be a median of ΔABC. Let these medians meet at G. Now in AC O_3 reflects into N_3. Thus O_3N_3 has AC as right-bisector. Not only that: AC has O_3N_3 as right-bisector. (O_3 is equidistant from A and C.) Thus B' the midpoint of AC, is also the midpoint of O_3N_3. And X is the midpoint of O_1O_3. Thus $B'X$ is parallel to O_1N_3 and equal in length to one-half of it. Since O_1N_3 is equal and parallel to BO_2, BO_2 is parallel to and twice XB'. Thus Δ's $GB'X$ and GBO_2 are similar and have sides in the ratio 1/2. Thus $O_2G = 2 \cdot GX$. But O_2X is a median of $\Delta O_1O_2O_3$. Thus G is the centroid of $\Delta O_1O_2O_3$.

Similarly $GB = 2 \cdot GB'$, and BB' is a median of ΔABC. Thus G is the centroid of ΔABC. Similarly $BN_1O_3N_2$ is a parallelogram.

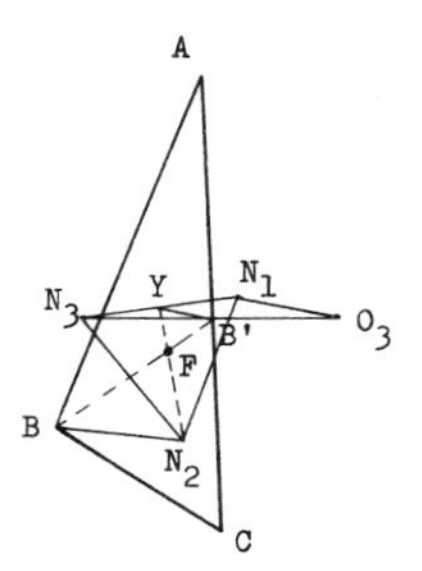

Let N_2Y be a median of $\Delta N_1N_2N_3$, and again let BB' be a median of ΔABC, meeting N_2Y in F. In $\Delta N_3O_3N_1$, YB' is parallel to and one-half of N_1O_3, which is equal and parallel to BN_2. Thus triangles YFB', FBN_2 are similar with sides as 1/2, etc., giving the result that F is the centroid of both ΔABC and $\Delta N_1N_2N_3$. Thus

$F = G$. Thus the outer and inner Napoleon triangles have the same centroid. (QED)

9. Choose three diagonals of H which form an equilateral triangle (of side 2). Place the pins at the midpoints of these diagonals. The proof is easy.

10. Mark corners as described in the problem at each vertex of H. Number these corners 1, 2, 3, 4, 5, 6, in order around H. The corners go together in three pairs of opposite corners. If one corner of a pair covers points of the set S, then the opposite corner cannot, lest S have diameter greater than 1. Thus, no matter how H covers S, some three corners cover no points of S, corners taken one from each of the three opposite pairs. For example, if 1, 3, 5 cover points of S, then corners 2, 4, 6 must be empty. Show that, no matter what three corners may be empty in this way, some two are separated by only one corner. H can then be turned so that these are the two corners which are removed. (QED)

12. The rotation through $+60°$ about B carries the diagonal AC into PQ (A into P, C into Q). The rotation about D through the same angle carries it into SR. Thus PQ and SR must be equal and parallel, implying $PSRQ$ is a parallelogram.

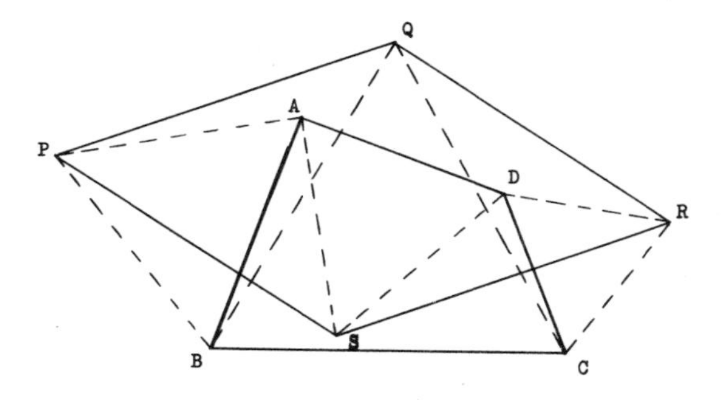

13. Reflect ΔABC in its sides to produce a regular hexagon around A, as shown. The curve in question, then, yields a closed curve because six reflections brings ΔABC back on to itself. The area in this inner curve is to be 1/2 the hexagon, or 3 times ΔABC, independent of the particular bisecting curve used. For minimal

length, the entire closed curve must be a circle. The solution is now easily completed.

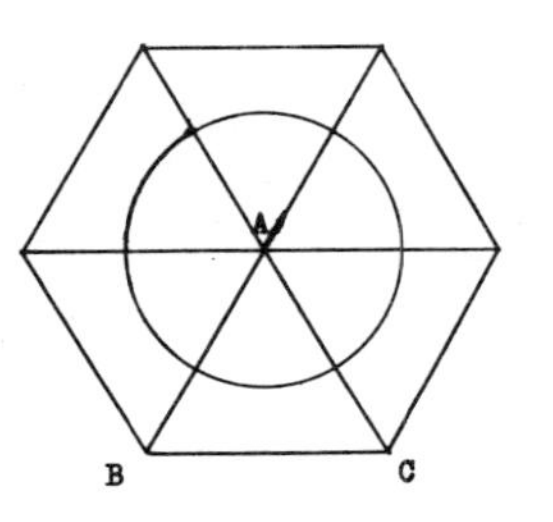

4. The Orchard Problem.

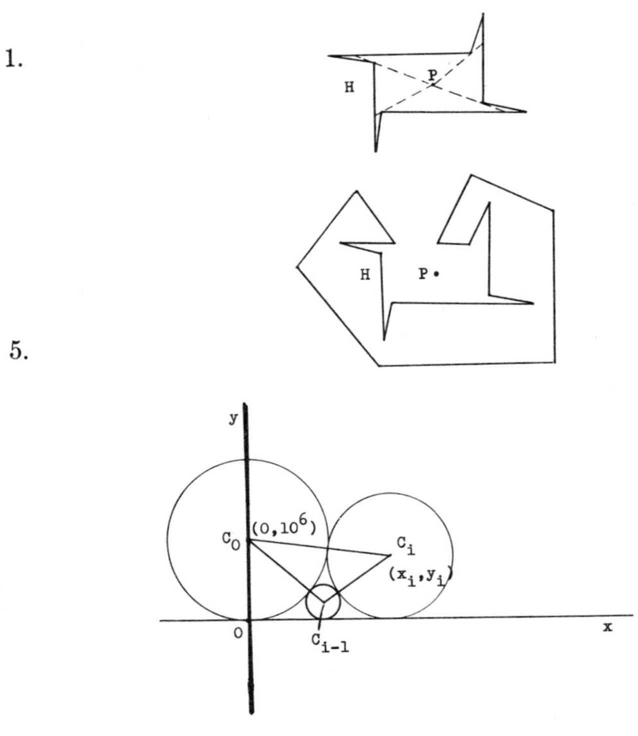

Let L be the x-axis and Z the origin of a cartesian frame of reference. Let circle C_i have center $c_i(x_i, y_i)$; this implies that the radius is y_i. Then, in general, we have $C_0C_i = 10^6 + y_i = \sqrt{x_i^2 + (y_i - 10^6)^2}$, giving $x_i^2 = 4\cdot 10^6 y_i$. Accordingly, $x_{i-1}^2 = 4\cdot 10^6 y_{i-1}$. Then

$$C_{i-1}C_i = y_i + y_{i-1} = \sqrt{(x_i - x_{i-1})^2 + (y_i - y_{i-1})^2}$$

yields

$$(x_i - x_{i-1})^2 = 4y_iy_{i-1}.$$

Then

$$(x_i - x_{i-1})^2 4\cdot 10^{12} = (4\cdot 10^6 y_i)(4\cdot 10^6 y_{i-1}) = x_i^2 x_{i-1}^2,$$

or

$$(x_i - x_{i-1})2\cdot 10^6 = x_i x_{i-1}.$$

This relation does not hold for $i = 1$, since $x_0 = 0$. In this case we have, by Pythagoras, $x_1^2 + (10^6 - 1)^2 = (10^6 + 1)^2$, giving $x_1 = 2\cdot 10^3$. By the general relation, this leads to $(x_2 - 2\cdot 10^3)\cdot 2\cdot 10^6 = x_2\cdot 2\cdot 10^3$, giving

$$x_2 = \frac{2\cdot 10^6}{10^3 - 1}.$$

Again we have

$$\left(x_3 - \frac{2\cdot 10^6}{10^3 - 1}\right)\cdot 2\cdot 10^6 = x_3 \frac{2\cdot 10^6}{10^3 - 1},$$

giving $x_3 = 2\cdot 10^6/(10^3 - 2)$. Noting that x_1 can be written $2\cdot 10^6/10^3$, we begin to suspect that, in general, $x_i = 2\cdot 10^6/(10^3 - i + 1)$. Let us attempt a proof of this by induction. For $i = 1$, $x_i = x_1 = 2\cdot 10^3$, while the formula gives $2\cdot 10^6/(10^3 - 1 + 1)$, which is the same. Suppose the formula is valid for $i = K - 1$. Then, using the recursion to solve for x_K, we obtain

$$(x_K - x_{K-1})\cdot 2\cdot 10^6 = \left(x_K - \frac{2\cdot 10^6}{10^3 - K + 2}\right)\cdot 2\cdot 10^6$$

$$= x_K\cdot \frac{2\cdot 10^6}{10^3 - K + 2}$$

giving

$$x_K = \frac{2 \cdot 10^6}{10^3 - K + 1},$$

which tells us our formula is again correct. Hence the formula is valid for all i.

Using again our first relation, $x_i^2 = 4 \cdot 10^6 y_i$, we obtain

$$4 \cdot 10^6 y_i = \frac{4 \cdot 10^{12}}{(10^3 - i + 1)^2}, \quad \text{giving} \quad y_i = \frac{10^6}{(10^3 - i + 1)^2}.$$

Now the process of drawing circles can continue until y_i exceeds 10^6. Then it stops. For $y_i > 10^6$ we must have the denominator $(10^3 - i + 1)^2 < 1$; i.e., $(10^3 - i + 1)^2 = 0$. This gives $i = 1001$.

Thus certainly no circle beyond C_{1001} can be drawn. But perhaps even C_{1001} is too big. This would be the case if y_{1000} happened to equal 10^6. Checking, we get

$$y_{1000} = \frac{10^6}{(10^3 - 1000 + 1)^2} = 10^6.$$

Thus, while in general one circle bigger than C_0 is possible, in the case at hand C_{1000} is the last circle. That is, after C_0 and C_1, 999 additional circles may be drawn.

7. The theorem of Schnirelman establishes the fact that some four points of our curve are at the vertices of a square. Since the curve is convex, the perimeter of the curve must be at least as great as the perimeter of the square. Thus the square has perimeter less than four, giving it a side less than unity.

Thus, centering the inscribed square on the unit square of the lattice (center on center) with its sides parallel to those of the unit square puts the curve so that it covers no lattice point. Since its side is less than 1, it lies strictly inside the unit square. Because the curve is convex, no point of it can lie outside the two strips formed by extending the sides of the inscribed square, thus it cannot cover a lattice point (e.g., if P belonged to our curve, then so would all ΔPAC, meaning B is not a boundary point).

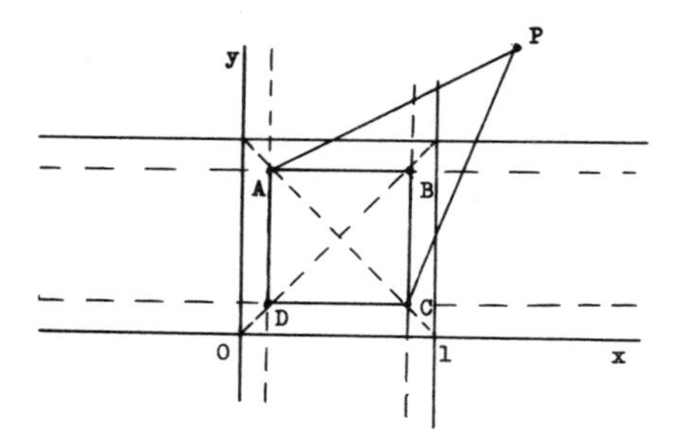

5. Δ-Curves.

2. Let PW and PV be perpendicular to AB, AC, respectively. Join PA, PU, PX. Now WV subtends a 60-degree angle at both A and X, implying $AXWV$ is cyclic. Because of the right angles, the circle on AP as diameter goes through W and V and A. Thus X, W, P, V, A are concyclic. Now $\angle XAW = \angle XVW = 60°$, making $\angle XAC = 120°$. Thus XA and CB are parallel. But because AP is a diameter, XP is perpendicular to AX. Thus XP is perpendicular to BC. On the generalization of exercise 1, we have UP perpendicular to BC. Thus XPU is straight and perpendicular to BC. Thus XU equals the distance between the parallels XA and BC, which is h.

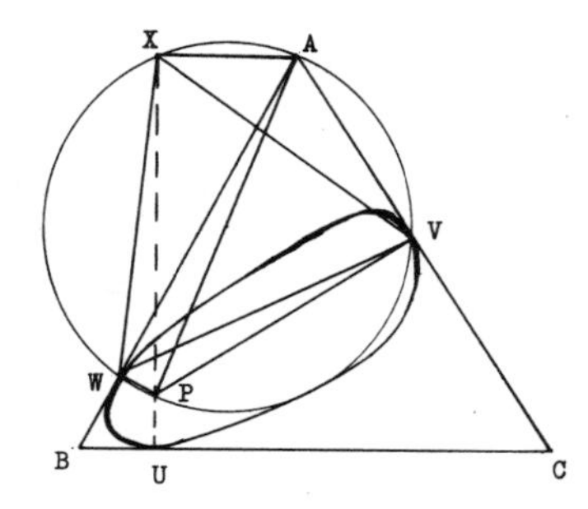

6. Let m_1, m_2 denote any pair of parallel supporting lines, and suppose A and B denote a point of contact on each. (If there is a

choice for A or B, any point of contact will do.) Let n_1, n_2 denote the pair of parallel supporting lines in the direction of AB. Let K and L denote a pair of points of contact on n_1, n_2.

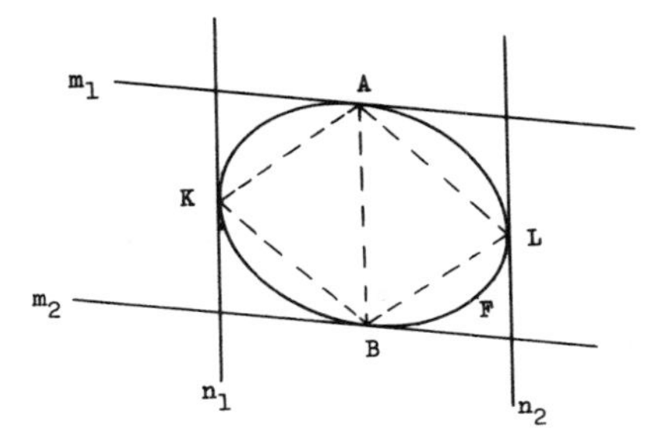

Now the four supporting lines enclose the given figure F in a parallelogram which is divided into two smaller parallelograms by AB. The triangles KAB and ABL are 1/2 their respective parallelograms, implying that the circumscribing parallelogram is twice $KBLA$, which is no greater than twice F ($KBLA$ is contained in F because F is convex).

6. It's Combinatorics that Counts!

6. A solution follows from the fact that the number of dominoes which cross a vertical or horizontal ruling of the board is always even. This is easily proven indirectly. A ruling crossed by an odd number of dominoes would determine a tiled region K which con-

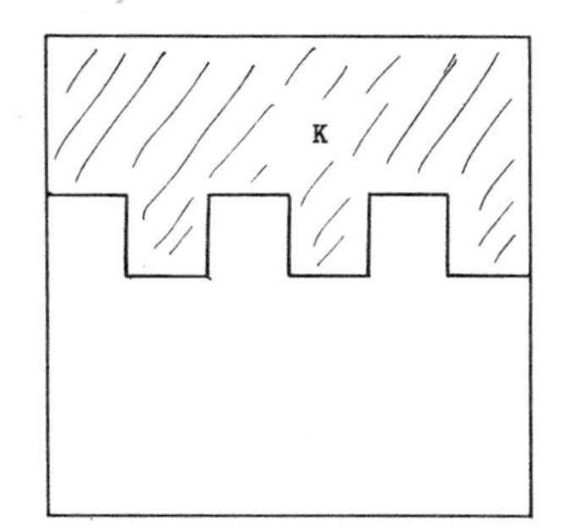

tains an odd number of squares. But this is impossible. Being completely tiled with a whole number of dominoes, K must have an even number of squares.

Now suppose that there is no ruling uncrossed by dominoes. Since the number of dominoes across a ruling is even, each ruling must be crossed by at least two dominoes. However, a domino can cross only one ruling. The ten rulings, then, require at least twenty different dominoes to cross them. But a 6×6 board has room for only $6^2/2 = 18$ dominoes.

7.

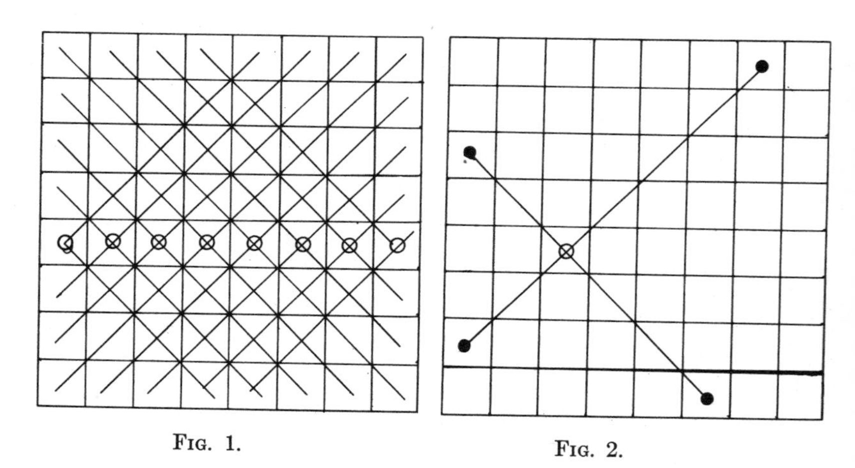

FIG. 1. FIG. 2.

Place a bishop in each square of the fourth (or fifth) rank. Figure 1 shows that 8 bishops will do. We need show that less than 8 will not do.

Figure 2 shows that no bishop can cover more than 4 of the squares of its color around the outside. By a direct count, we see that there are 14 white and 14 black squares around the outside. Thus at least 4 bishops of each color are required. (We assume a bishop controls the square it occupies.)

8. Let the number of diagonal moves be r. Then the number of horizontal and the number of vertical moves is $7 - r$. The total

number of moves is $14 - r$. The number of ways of arranging three groups of moves (r alike of one kind, $7 - r$ alike of another, and $7 - r$ alike of a third kind) is

$$\frac{(14 - r)!}{(7 - r)!\,(7 - r)!\,r!}.$$

Adding as r goes from 0 to 7, we get 48, 639.

7. The Kozyrev-Grinberg Theory of Hamiltonian Circuits.

1. (Indirect) Suppose y is not used but x is used. Then: edges 1, 2, and 3, 4 are forced; this eliminates a, b, lest a premature triangle is formed. Thus 5, 6 are forced, making the valence of the circuit at the vertex P equal to 3; impossible.

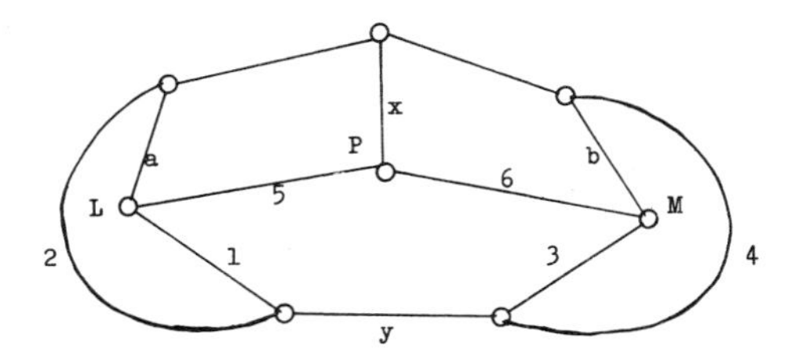

2. If two of x, y, z are not used, (suppose the two not used are y and z) then p, q, and r are forced, closing a premature triangle; impossible. Thus at least two of x, y, z must be used. However, all 3 cannot be used, for in that case, (a) if r is used, then p, q are not used, meaning two of the three edges at vertex K are not used; impossible; (b) thus r cannot be used; thus p and q are both used; similarly a, b and m, n are used, closing a premature 9-gon. Thus exactly two of x, y, z must be used. If a Hamiltonian circuit contains both p and q, then r cannot be used, forcing y and z to be used. But exactly two of x, y, z must be used, eliminating x.

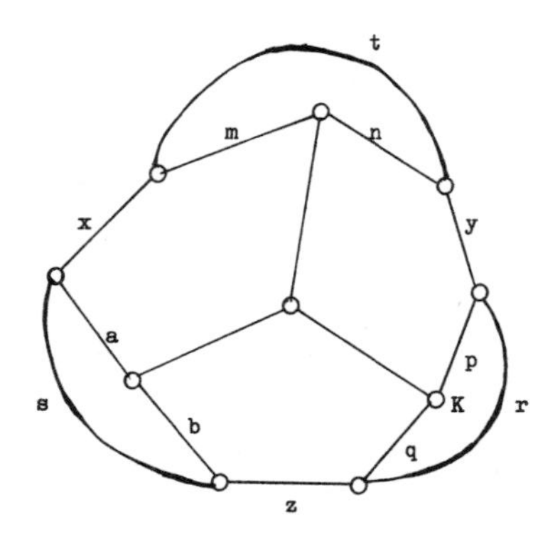

3. The Kozyrev-Grinberg relation gives $2(f_4 - f_4') + 3(f_5 - f_5') = 0$, f_4 and f_4' cannot be equal since their sum is 5, an odd number. Thus $f_4 - f_4'$ is not zero, implying $f_5 - f_5'$ is not zero. Since there are only two pentagons, then $f_5 - f_5'$ is either 2 or -2. Thus both pentagons lie on the same side of any Hamiltonian circuit.

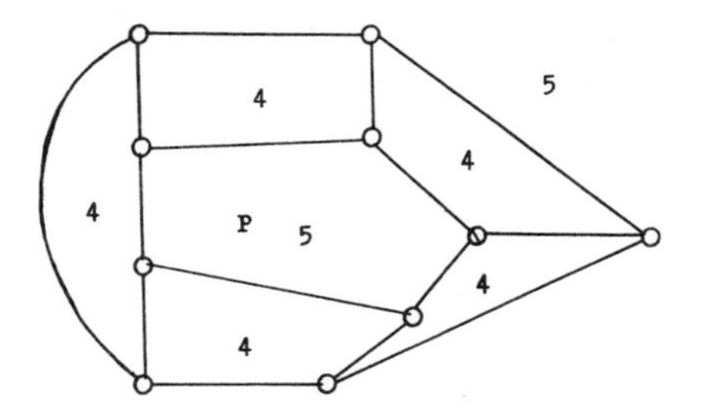

Since the infinite pentagon lies outside, so does P. Then

$$f_5 - f_5' = -2,$$

giving

$$2(f_4 - f_4') - 6 = 0 \quad \text{or} \quad f_4 - f_4' = 3.$$

Since there are 5 4-gons, the only choice is $f_4 = 4$, $f_4' = 1$. Since the edges of P all separate P from a 4-gon, exactly 4 of its edges belong to a Hamiltonian circuit and one does not.

4. (a) We have only one 4-gon; thus, $-2 + 3(f_5 - f_5') + 6(f_8 - f_8') = 0$, implying 3 divides 2; impossible.

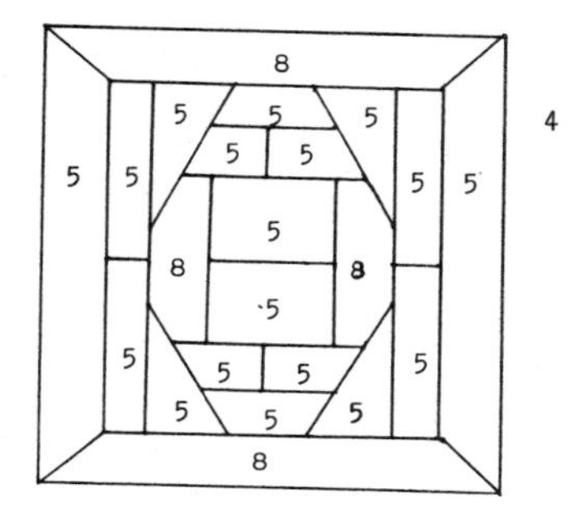

(b) Here we have 1 4-gon, 2 11-gons, and a bunch of 5-gons; thus, $\pm 2 + 3(f_5 - f_5') + 9(f_{11} - f_{11}') = 0$, implying 3 divides 2; impossible.

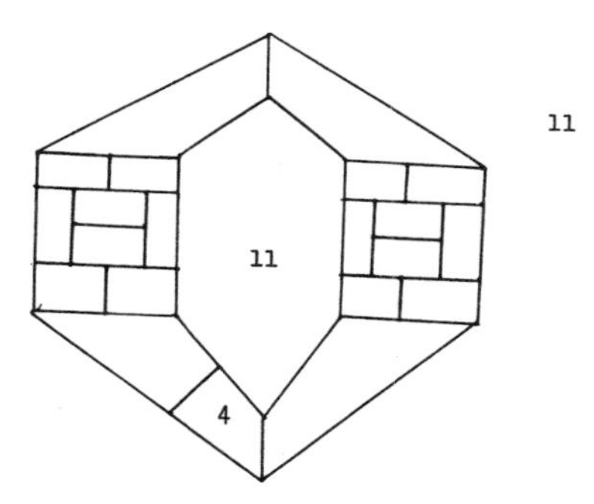

(c) This is the same graph as (b), except an extra edge changes things to 2 6-gons, 1 9-gon, 1 11-gon, and a bunch of 5-gons. Thus $3(f_5 - f_5') + 4(f_6 - f_6') \pm 7 - 9 = 0$.

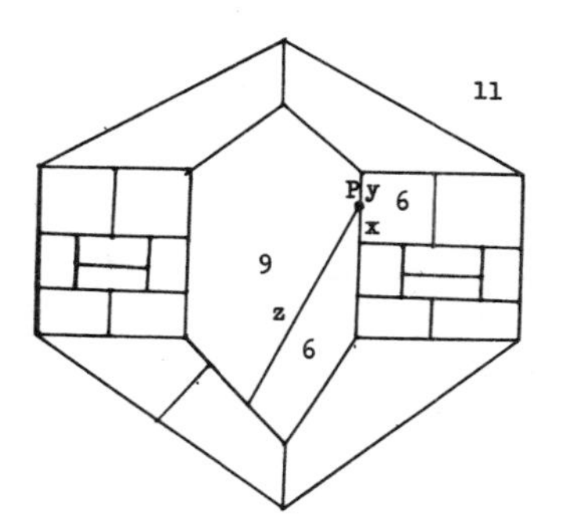

Now $f_6 - f_6' = 0$, 2 or -2. If 0, then 3 divides 7; impossible. Thus it is either 2 or -2. Then $3(f_5 - f_5') \pm 8 \pm 7 = 9$. Since $f_6 - f_6' = 2$ or -2, both 6-gons lie on the same side of any Hamiltonian circuit. This means, at vertex P, that edge x cannot be used, forcing the use of both y and z. This means the 9-gon is on the side of the Hamiltonian circuit opposite to that of the 6-gons. Thus the signs of the 8 and the 7 in the above equation must be different. Thus we have $3(f_5 - f_5') \pm 1 = 9$, implying 3 divides 1; impossible.

5. Let Q denote the graph in question. Since the valence at N is only 2, both edges at N, as well as LM, must belong to the Hamiltonian circuit.

Add two vertices A and B on LM and a vertex C on an edge at N to produce the graph Q'. Then a Hamiltonian circuit of Q which contains LM and the edges at N gives a Hamiltonian circuit of Q'. The faces of Q' are 4-gons, 5-gons, and 11-gons. Kozyrev-Grinberg implies

$$2(f_4 - f_4') + 3(f_5 - f_5') + 9(f_{11} - f_{11}') = 0,$$

which means that $f_4 - f_4'$ must be divisible by 3. However, there is only one 4-gon, making $f_4 - f_4'$ either 1 or -1.

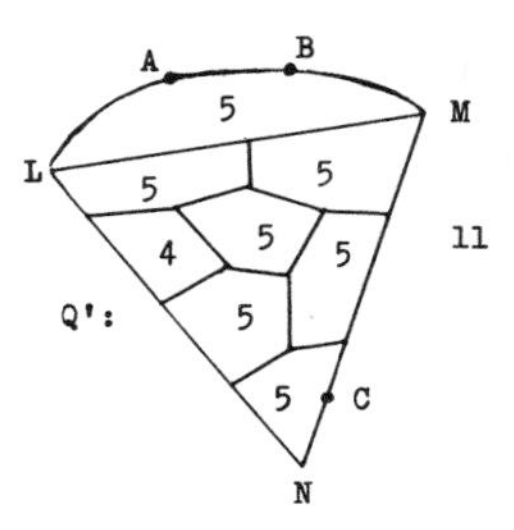

8. Morley's Theorem.

1. $S = 180° - 2b - 2c$, $\frac{1}{2}S = 90° - b - c$. Thus $\angle BPR = 180° - b - 30° - \frac{1}{2}S = 150° - b - 90° + b + c = 60° + c$. Similarly, $\angle CPQ = 60° + b$.

3. These lines bisect the angles of equilateral ΔPQR.

5. Let the exterior angles be $3a$, $3b$, and $3c$. Then $3a + 3b + 3c = 360°$, or $a + b + c = 120°$. Thus at least one of a, b, c is 40 degrees or more; say a. Then $b + c \leqq 80°$ and $2b + 2c \leqq 160°$, implying A and S are on opposite sides of BC. Now $S = 180° - 2b - 2c = 180° - (240° - 2a) = 2a - 60°$, and $\frac{1}{2}S = a - 30°$. (Note that $3a < 180°$, giving $a < 60°$, making $a - 30° < 30°$.)

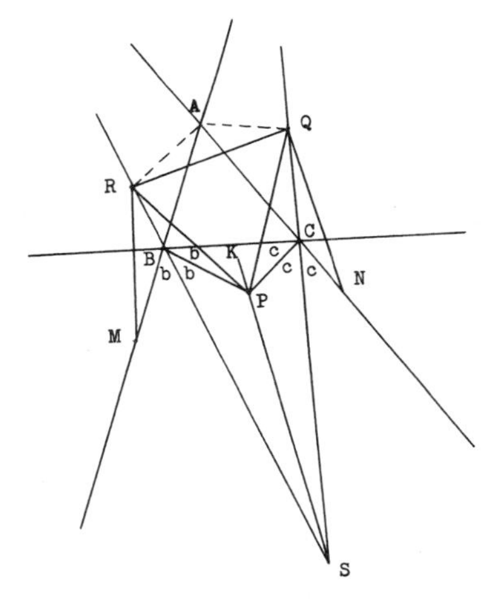

Constructing angles KPR and KPQ equal to 30 degrees, then, gives points R and Q on SB and SC extended, respectively. Now Δ's RPS, QPS are congruent, because PS is the third angle-bisector of ΔBCS (allowing AAS to apply). Thus $RP = QP$. Since $\angle RPQ = 60°$, then ΔPQR is equilateral. It remains to show that RA and QA trisect the exterior angles at A. Reflect P in BS and CS to give M and N. Then $MR = RP = PQ = QN$ and which also equals QR. Now angle $MRQ = 60° + 2(\angle BRP) = 60° + 2(30° - \frac{1}{2}S) = 120° - S = 180° - 2a$. And $\angle RQN$ is the same size. As before, $MRQN$ is cyclic. And the radius to R bisects $\angle MRQ$; i.e., the angle subtended at the center of this circle by MR is

$$180° - 2(\tfrac{1}{2} \angle MRQ) = 180° - (180° - 2a) = 2a.$$

Thus, at the circumference, MR subtends an angle of a. Similarly for QR, QN. Consequently, MN subtends at the center an angle of $6a$. Since a is at least $40°$, this is a reflex angle. The non-reflex angle subtended at the center by MN is then $360° - 6a$. The angle subtended at the circumference of the circle, in the segment containing the center, is $180° - 3a$. But this is the angle A of the given ΔABC. Thus the vertex A lies on this segment of this circle. So do R and Q lie on this same segment. Thus $\angle RAM = a$, and $\angle QAN = a$, implying the desired conclusion.

9. A Problem in Combinatorics.

1. Let the number of triangles be t. Then the sum of the degree measures of all the angles in the figure is

$$180t = 360m + (n - 2)\cdot 180$$

giving

$$t = 2m + n - 2.$$

2. (Indirect) Suppose all faces have different numbers of edges. The one with the least number of edges has at least 3 edges; the "next greatest" has at least 4 edges, etc., down to the very greatest, i.e., the Fth face in this ordering, must have at least $F + 2$ edges. But each edge around this largest face denotes a bordering face. Thus there must be at least $F + 3$ faces; contradiction—there are F faces.

4. An $a \times b$ rectangle has $9 - a$ positions across the board and $9 - b$ positions down the board, giving a total of $(9 - a)(9 - b)$ places for it on the board. Thus the total number of rectangles is $\Sigma\,(9 - a)(9 - b)$ where a, b range over 1, 2, 3, 4, 5, 6, 7, 8.

For squares we merely have just those values for which $a = b$. For $a = 1$: we get $8[8 + 7 + \cdots + 1] = 8\cdot 36$ (giving b successively the values $1, 2, \ldots, 8$). For $a = 2$: we get 7[same quantity] $= 7\cdot 36$, etc. The total is $[8 + 7 + \cdots + 1]\cdot 36 = 36\cdot 36 = 36^2 = 1296$. For squares, we have only the terms $8\cdot 8 + 7\cdot 7 + \cdots + 1\cdot 1 = 204$. Similarly, for an $n \times n$ board, we have $[n(n + 1)/2]^2$, which is merely the sum of the cubes of the first n natural numbers;

for the squares we have simply the sum of the squares of the first n natural numbers.

8. Every polygon with central symmetry has opposite each side another side which is equal and parallel to it. Such polygons, then, have an even number of sides. Also any such quadrilateral must be a parallelogram. Thus, we need to show that there are at least six faces that are quadrilaterals. Let F_n denote the number of faces with n sides. Then, for n odd, we have $F_n = 0$. Let V_n denote the number of vertices with valence n. Then $F = F_3 + F_4 + F_5 + \cdots$, and $V = V_3 + V_4 + V_5 + \cdots$; $F = F_4 + F_6 + F_8 + \cdots$, since $F_{2k+1} = 0$. We want to show that $F_4 \geqq 6$. Counting the edges around the faces and then around the vertices, we get

$$2E = 4F_4 + 6F_6 + 8F_8 + \cdots, \quad 2E = 3V_3 + 4V_4 + 5V_5 + \cdots.$$

Now $\quad V - E + F = 2, \quad$ or $\quad 6V - 6E + 6F = 12,$

or $\quad 6V - 2E + 6F = 12 + 4E.$

Thus
$$6(V_3 + V_4 + V_5 + \cdots) - (4F_4 + 6F_6 + 8F_8 + \cdots)$$
$$+ 6(F_4 + F_6 + F_8 + \cdots)$$
$$= 12 + 2(3V_3 + 4V_4 + 5V_5 + \cdots),$$
$$2F_4 - 2F_8 - 4F_{10} - 6F_{12} - \cdots$$
$$= 12 + 2V_4 + 4V_5 + 6V_6 + \cdots,$$
giving $F_4 = 6 + (V_4 + 2V_5 + 3V_6 + \cdots)$
$$+ (F_8 + 2F_{10} + 3F_{12} + \cdots),$$
implying $F_4 \geqq 6$.

9. Let n denote the number of pentagonal faces. Then the number of hexagonal faces is $F - n$. Counting edges, we get $2E = 5n + 6(F - n)$. This gives $n = 6F - 2E$. Now $V - E + F = 2$, and, because the valence at every vertex is 3, the average is 3, giving $2E = 3V$. Thus $\frac{2}{3}E - E + F = 2$, which is $6F - 2E = 12$. Thus there are 12 pentagonal faces.

10. Multiply-Perfect, Superabundant, and Practical Numbers.

4. Suppose $\sigma(n)$ is odd. Express n in the form $2^a k$ where k is odd. Then $\sigma(n) = \sigma(2^a k) = \sigma(2^a)\cdot\sigma(k) = (2^{a+1} - 1)\sigma(k)$ is odd. Since $2^{a+1} - 1$ is odd, this implies $\sigma(k)$ is odd. However, since k is odd, all its divisors are odd, too. In order to make $\sigma(k)$ odd, then, there must be an odd number of divisors of k. Hence k must be a square, say m^2. Then $n = 2^a \cdot m^2$. If a is even, then n, too, is a square; if odd, then n is twice a square.

Conversely, if n is a square (t^2) or twice a square ($2t^2$) we can argue as follows:

Suppose $t = 2^b \cdot p_2^{a_2} \cdot p_3^{a_3} \cdots p_v^{a_v}$ in its prime decomposition. Then $t^2 = 2^{2b} \cdot p_2^{2a_2} \cdot p_3^{2a_3} \cdots p_v^{2a_v}$, and $2t^2 = 2^{2b+1} \cdot p_2^{2a_2} \cdot p_3^{2a_3} \cdots p_v^{2a_v}$. Then $\sigma(n) = \sigma(t^2) = (2^{2b+1} - 1)\cdot\sigma(p_2^{2a_2})\cdot\sigma(p_3^{2a_3})\cdots\sigma(p_v^{2a_v})$, or $\sigma(n) = \sigma(2t^2) = (2^{2b+2} - 1)\cdot\sigma(p_2^{2a_2})\cdot\sigma(p_3^{2a_3})\cdots\sigma(p_v^{2a_v})$. Here every p_i, $i = 2, 3, \ldots, v$, is an odd prime. Thus every divisor p_i^m of $p_i^{2a_i}$ is odd. Since the index $2a_i$ is even, then the value of $\sigma(p_i^{2a_i}) = 1 + p_i + p_i^2 + \cdots + p_i^{2a_i}$ is also odd. But $2^{2b+1} - 1$ and $2^{2b+2} - 1$ are odd. Thus $\sigma(n)$ is odd in all cases. Thus a natural number n has $\sigma(n)$ odd if and only if n is a square or twice a square.

5. Let $n = p_1^{a_1} \cdot p_2^{a_2} \cdots p_k^{a_k}$ denote an odd perfect number; then all the primes p_i are odd. Because n is perfect, we have

$$\sigma(n) = \sigma(p_1^{a_1})\cdot\sigma(p_2^{a_2})\cdots\sigma(p_k^{a_k}) = 2\cdot p_1^{a_1}\cdot p_2^{a_2}\cdots p_k^{a_k}.$$

In the prime decomposition of the right-hand side, a single 2 occurs. The same follows for the left-hand side. Thus one of the $\sigma(p_i^{a_i})$ contains a single 2 in its prime decomposition, while the rest are odd. Since only one 2 occurs, this $\sigma(p_i^{a_i})$ is twice an odd number. For definiteness, suppose $\sigma(p_1^{a_1}) = 2(2q + 1) = 4q + 2$. Since $\sigma(p_i^{a_i})$ is odd for $i = 2, 3, \ldots, k$, we must have $p_i^{a_i}$ either a square or twice a square. Since $p_i^{a_i}$ is odd, it cannot be twice a square. Thus each $p_i^{a_i}$ is a square, implying that

$$p_2^{a_2} \cdot p_3^{a_3} \cdots p_k^{a_k} = Q^2, \qquad \text{a square.}$$

Now we show that $a_1 = 4a + 1$ for some integer a. We have

$\sigma(p_1^{a_1}) = 4q + 2$, which is congruent to 2 mod 4. Then

$$1 + p_1 + p_1^2 + \cdots + p_1^{a_1} \equiv 2 \pmod 4.$$

Since $\sigma(p_1^{a_1})$ is an even number, it must be that a_1 is odd. Hence a_1 is either of the form $4t + 1$ or $4t + 3$. We suppose $a_1 = 4t + 3$ and deduce a contradiction. We see, then, that $\sigma(p_1^{4t+3})$ contains $4t + 4$ terms, each of which is a power of an odd prime (p_1).

Since p_1 is odd, then $p_1 \equiv 1 \pmod 4$ or $p_1 \equiv 3 \pmod 4$.

In the case $p_1 \equiv 1 \pmod 4$, we have every power of $p_1 \equiv 1 \pmod 4$, implying that $\sigma(p_1^{4t+3}) \equiv (4t + 4)\cdot 1 \equiv 0 \pmod 4$, a contradiction. If $p_1 \equiv 3 \pmod 4$, then even powers of p_1 are $\equiv 1 \pmod 4$ while odd powers are $\equiv 3 \pmod 4$; i.e., the sum of an odd and an even power of p_1 is $\equiv 0 \pmod 4$. Then

$$\sigma(p_1^{4t+3}) = (1 + p_1) + (p_1^2 + p_1^3) + \cdots + (p_1^{4t+2} + p_1^{4t+3}) \equiv 0 \pmod 4,$$

a contradiction.

11. Circles, Squares, and Lattice Points.

1. I: If $\left|\frac{a+b}{2}\right| + \left|\frac{a-b}{2}\right| < c$, then

$$|a| = \left|\frac{a+b}{2} + \frac{a-b}{2}\right| \leqq \left|\frac{a+b}{2}\right| + \left|\frac{a-b}{2}\right| < c;$$

and

$$|b| = \left|\frac{a+b}{2} - \frac{a-b}{2}\right|$$

$$\leqq \left|\frac{a+b}{2}\right| + \left|-\frac{a-b}{2}\right| = \left|\frac{a+b}{2}\right| + \left|\frac{a-b}{2}\right| < c.$$

II: If $|a| < c$ and $|b| < c$, then if one of $(a + b)/2$, $(a - b)/2$

is zero, the other is a. Consequently,

$$\left|\frac{a+b}{2}\right| + \left|\frac{a-b}{2}\right| = |a| < c.$$

For nonzero $(a + b)/2$, $(a - b)/2$ we have:

If they have the same sign, then

$$\left|\frac{a+b}{2}\right| + \left|\frac{a-b}{2}\right| = \left|\frac{a+b}{2} + \frac{a-b}{2}\right| = |a| < c.$$

If they have different signs, then

$$\left|\frac{a+b}{2}\right| + \left|\frac{a-b}{2}\right| = \left|\frac{a+b}{2} - \frac{a-b}{2}\right| = |b| < c. \qquad \text{(QED)}$$

2. We show that no sphere with center $(\sqrt{2}, \sqrt{3}, \sqrt{5})$ has more than one lattice point on its surface. The contrary gives an integral solution to

$$(a - \sqrt{2})^2 + (b - \sqrt{3})^2 + (c - \sqrt{5})^2$$
$$= (d - \sqrt{2})^2 + (e - \sqrt{3})^2 + (f - \sqrt{5})^2,$$

$$\sqrt{2}(2d - 2a) + \sqrt{3}(2e - 2b) + \sqrt{5}(2f - 2c)$$
$$= d^2 + e^2 + f^2 - a^2 - b^2 - c^2.$$

We prove later that if $p\sqrt{2} + q\sqrt{3} + r\sqrt{5} =$ a rational number, then $p = q = r = 0$. This gives $a = d$, $b = e$, $c = f$, implying the lattice points (a, b, c), (d, e, f) are the same, contradiction. (QED)

Thus no two lattice points are the same distance from $(\sqrt{2}, \sqrt{3}, \sqrt{5})$, giving an ordering $p_1, p_2, \ldots, p_n, \ldots$ according to their increasing distance from $(\sqrt{2}, \sqrt{3}, \sqrt{5})$. Consequently, the sphere with center $(\sqrt{2}, \sqrt{3}, \sqrt{5})$ through p_{n+1} contains exactly n lattice points $p_1, p_2, \ldots, p_n$. (QED)

Proof of Lemma: If $p\sqrt{2} + q\sqrt{3} + r\sqrt{5} =$ *a rational number, then* $p = q = r = 0$ (p, q, r rational).

Let $p\sqrt{2} + q\sqrt{3} + r\sqrt{5} = k$, a rational number. Then $p\sqrt{2} + q\sqrt{3} = k - r\sqrt{5}$, and $2p^2 + 2pq\sqrt{6} + 3q^2 = k^2 - 2kr\sqrt{5} + 5r^2$. Thus $2pq\sqrt{6} + 2kr\sqrt{5} = k^2 + 5r^2 - 2p^2 - 3q^2$, which is rational. Let this be denoted $a\sqrt{6} + b\sqrt{5} = c$. Here a, b, c are rational. Then $6a^2 + 5b^2 - c^2 = -2ab\sqrt{30}$.

Since $\sqrt{30}$ is irrational, we must have $2ab = 0$ and $6a^2 + 5b^2 - c = 0$. Thus a or b is zero. Suppose $a = 0$. Then $b\sqrt{5} = c$, implying $b = c = 0$. If $b = 0$, then $a\sqrt{6} = c$, implying $a = c = 0$. In any case $a = b = c = 0$.

Thus $2pq = 2kr = k^2 + 5r^2 - 2p^2 - 3q^2 = 0$. Accordingly, either p or q is zero. Suppose $p = 0$. Then $q\sqrt{3} + r\sqrt{5} = k$, and as with $a\sqrt{6} + b\sqrt{5} = c$, we get $q = r = k = 0$. Similarly for $q = 0$. Thus in all cases $p = q = r = 0$. (QED)

3. We need only show that for each rational point (x, y) there exist two lattice points which are the same distance from it. If the circle with center (x, y) through these two lattice points contains n lattice points in its interior, then every larger circle with center (x, y) contains at least $n + 2$ lattice points, and no circle with center (x, y) can contain exactly $n + 1$ lattice points.

Let $x = p/q$ and $y = r/q$, when brought to a positive common denominator q. (Then p and r may be positive or negative.) Then the lattice points $(r, -p)$ and $(-r, p)$ are the same distance from (x, y), which is the point $(p/q, r/q)$. This is because

$$\left(r - \frac{p}{q}\right)^2 + \left(-p - \frac{r}{q}\right)^2 = \left(-r - \frac{p}{q}\right)^2 + \left(p - \frac{r}{q}\right)^2.$$

(QED)

12. Recursion.

2. Let t denote the number of triangles. Then the sum of the degree measures of all the angles in the figure is

$$180t = (n - 2)180,$$

because all the angles are at the vertices of the n-gon. Hence $t = n - 2$.

Let d denote the number of diagonals. Counting each triangle at 3 edges, the total number of edges is $3n - 6$. However, this counts each diagonal twice and each side of the n-gon once. Thus

$$3n - 6 = 2d + n, \qquad \text{giving} \quad d = n - 3.$$

3. Let x denote the probability of the fly getting stuck at G if it begins at H. Similarly let y and z denote the probabilities for the vertices E and D, respectively. By symmetry, the probabilities at B and C, respectively, are x and y. Let the required probability be p. Then we have

$$p = \tfrac{1}{3}x + \tfrac{1}{3}x + \tfrac{1}{3}z; \qquad x = \tfrac{1}{3}p + \tfrac{1}{3}y + \tfrac{1}{3}\cdot 1$$

$$y = \tfrac{1}{3}x + \tfrac{1}{3}z + \tfrac{1}{3}\cdot 0; \qquad z = \tfrac{1}{3}p + \tfrac{1}{3}y + \tfrac{1}{3}y.$$

Solving these gives $p = \frac{4}{7}$.

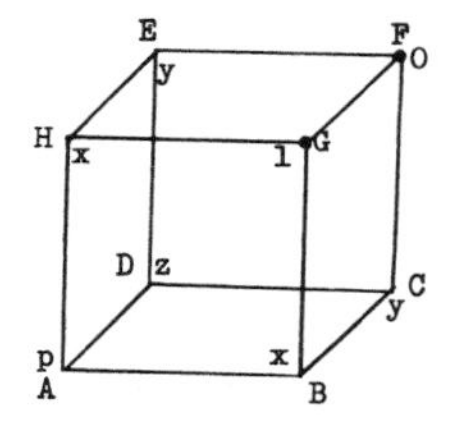

Considering getting stuck at F, we have similar equations from which $p = 3/7$. Thus the probability that the fly does not get stuck at all is

$$1 - (\tfrac{4}{7} + \tfrac{3}{7}) = 0.$$

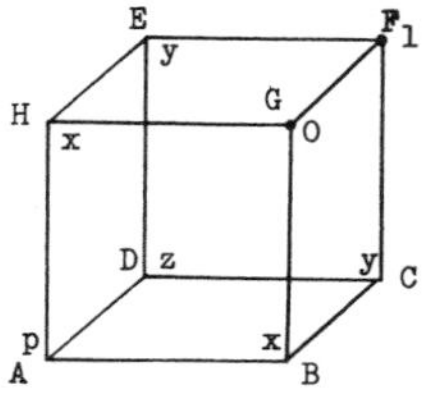

4. It easily turns out that $(1 - x - x^2)F(x) = 1$. Let the roots of $x^2 + x - 1 = 0$ be y and z. Then $x^2 + x - 1 = (x - y)(x - z)$, giving $1 - x - x^2 = -(y - x)(z - x)$. Let $F(x) = 1/(1 - x - x^2)$ $= A/(y - x) + B/(z - x)$. Then $1 = -A(z - x) - B(y - x)$. Equating coefficients we get $A + B = 0$ and $-Az - By = 1$. Thus $B = -A$ and $B(z - y) = 1$, or $B = 1/(z - y)$, and $A =$

$-1/(z - y)$. Then

$$F(x) = \frac{1}{z - y}\left[-\frac{1}{y - x} + \frac{1}{z - x}\right].$$

Now the actual values of y and z are $y = (-1 + \sqrt{5})/2$, $z = (-1 - \sqrt{5})/2$. Hence $z - y = -\sqrt{5}$. Thus

$$F(x) = \frac{1}{\sqrt{5}}\left[\frac{1}{y - x} - \frac{1}{z - x}\right] = \frac{1}{\sqrt{5}}[(y - x)^{-1} - (z - x)^{-1}]$$

$$= \frac{1}{\sqrt{5}}\left[y^{-1}\left(1 - \frac{x}{y}\right)^{-1} - z^{-1}\left(1 - \frac{x}{z}\right)^{-1}\right].$$

Now the coefficient of all terms in an expansion $(1 - w)^{-1}$ is 1. Also we observe that the product of the roots y and z is -1. Thus y and z are negative reciprocals. Hence

$$F(x) = \frac{1}{\sqrt{5}}\left[-z\left(1 - \frac{x}{y}\right)^{-1} + y\left(1 - \frac{x}{z}\right)^{-1}\right].$$

Equating coefficients of x^n, we get

$$f_{n+1} = \frac{1}{\sqrt{5}}\left[-z\left(\frac{1}{y}\right)^n + y\left(\frac{1}{z}\right)^n\right] = \frac{1}{\sqrt{5}}[(-z)^{n+1} - (-y)^{n+1}]$$

or

$$f_n = \frac{1}{\sqrt{5}}[(-z)^n - (-y)^n],$$

$$f_n = \frac{1}{\sqrt{5}}\left[\left(\frac{1 + \sqrt{5}}{2}\right)^n - \left(\frac{1 - \sqrt{5}}{2}\right)^n\right].$$

We take the opportunity of including here

The Neatest Derivation of Binet's Formula

Lemma. *If $x^2 = x + 1$, then $x^n = f_n x + f_{n-1}$ for $n = 2, 3, 4, \ldots$.*

Proof. (induction). For $n = 2$, $x^2 = x + 1 = f_2 x + f_1$, as required. Suppose $x^n = f_n x + f_{n-1}$. Then $x^{n+1} = x \cdot x^n = f_n x^2 + f_{n-1} x$ giving

$x^{n+1} = f_n(x + 1) + f_{n-1}x = (f_n + f_{n-1})x + f_n = f_{n+1}x + f_n$. The lemma follows by induction.

Now let the roots of $x^2 - x - 1 = 0$ be y and z. Then for $n = 2, 3, \ldots$, we have $y^n = f_n y + f_{n-1}$ and also $z^n = f_n z + f_{n-1}$, using the lemma. Subtracting gives $y^n - z^n = f_n(y - z)$, or $f_n = (y^n - z^n)/(y - z)$. But y and z are just $(1 + \sqrt{5})/2$ and $(1 - \sqrt{5})/2$. Thus

$$f_n = \frac{1}{\sqrt{5}}\left[\left(\frac{1+\sqrt{5}}{2}\right)^n - \left(\frac{1-\sqrt{5}}{2}\right)^n\right].$$

5. Each of the numbers 1, 2, 3, 4, 5, 6 is equally likely to come up on a roll of a die. In 6 rolls, then, we expect each of them to come up once, giving a total of 21 in 6 rolls, i.e., 7 in 2 rolls. Thus every 2 rolls we expect to advance the running total by 7. Thus we expect the running total to take a value twice in every seven values. The probability that the random integer is one of the values of the total, then, is 2/7.

6. Let c_n denote the number of ways for $2n$ points. Consider now $2n + 2$ points, for which the number of ways is c_{n+1}.

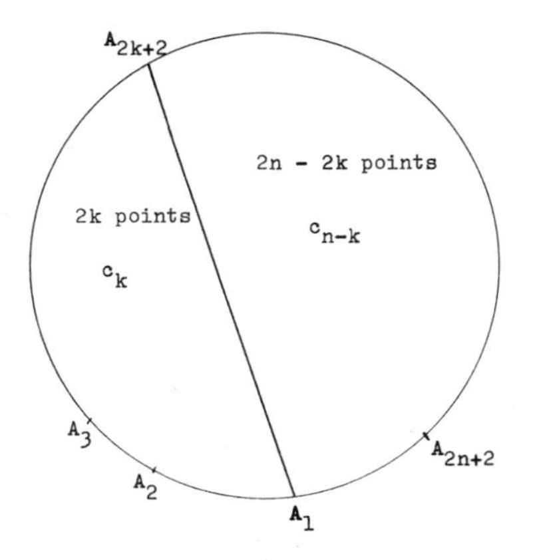

Order the points $A_1, A_2, \ldots, A_{2n+2}$ around the circle. Suppose A_1 is joined to A_t. If $t = 3$, we see that any chord to A_2 must intersect A_1A_3. By induction, we see that a similar impossible situation develops if t is any odd value. Thus t is even, say $2k + 2$.

The chord A_1A_{2k+2} separates the points into a group of $2k$ and a group of $2n - 2k$. Any of the c_k ways of pairing the first group combined with any of the c_{n-k} ways of pairing the second group provide a way of pairing the $2n + 2$ points. Thus

$$c_{n+1} = \sum_{k=0}^{k=n} c_k c_{n-k}, \qquad \text{where } c_0 = 1.$$

As in the essay, then, we obtain $c_n = (2n)!/n!(n + 1)!$.

13. Poulet, Super-Poulet, and Related Numbers.

4. By trial, one gets the answer $n = 6$, for the first part. For $n = 1, 2, 3, 4, 5, 6$ we get values $3^n - 3 = 0, 6, 24, 78, 240, 726$; $2^n - 2 = 0, 2, 6, 14, 30, 62$; n divides both for $n = 1, 2, 3, 5$. For $n = 4$, n does not divide $3^n - 3$; $n = 6$ divides $3^n - 3 = 726$ but not $2^n - 2 = 62$.

In the second part we reason as follows: since n does not divide $3^n - 3$, n cannot be a prime number (Fermat's Simple Theorem). Thus n is composite and divides $2^n - 2$, which makes it a pseudoprime. Thus we seek the least pseudoprime which does not divide $3^n - 3$. Since $n = 341$ is the least pseudoprime, we try it first. It is found not to divide $3^{341} - 3$ by showing that its factor 31 doesn't divide. We have, modulo 31, $3^3 = 27 \equiv -4$; $3^9 \equiv -64 \equiv -2$; $3^{54} \equiv 64 \equiv 2$; $3^{324} \equiv 64 \equiv 2$; Now $3^6 \equiv 16$, and $3^2 \equiv 9$, giving $3^8 \equiv 20$. Thus $3^{17} \equiv 3^9 \cdot 3^8 \equiv (-2) \cdot 20 \equiv -40 \equiv -9$. Hence $3^{341} \equiv 2(-9) \equiv -18$, giving $3^{341} - 3 \equiv -21$, not zero.

5. Every prime p divides $2^p - 2$, by Fermat's Simple Theorem. If p also divides $2^p + 1$, then it divides their difference, namely 3. The only prime to divide 3 is $p = 3$. And we find that 3 does divide $2^3 + 1 = 8 + 1 = 9$. Thus $p = 3$.

6. There are four kinds of natural numbers: 1, the composite numbers, the even prime 2, the odd primes. We consider these separately.

If $a = 1$: We seek a composite n such that n divides $1^n - 1 = 0$. Any composite will do, in particular the least, 4.
If *a is composite*: Here we may use $n = a$, because $a^a - a$ is divisible by a, and a is composite.
If *a is an odd prime*: Here we may use $n = 2a$. The dividend in this case is the even number $a^{2a} - a$ (both terms here are odd). Being even, it is divisible by 2. But it is also clearly divisible by a. Thus it is divisible by $n = 2a$ (since a is odd).
If $a = 2$: Here we may use any pseudoprime, say $n = 341$.

INDEX